Atmosphere-Biosphere Interactions: Toward a Better Understanding of the Ecological Consequences of Fossil Fuel Combustion

A Report Prepared by the
Committee on the Atmosphere and the Biosphere

Board on Agriculture and Renewable Resources
Commission on Natural Resources
National Research Council

NATIONAL ACADEMY PRESS
Washington, D.C.   1981

   This study was supported by the U.S. Environmental Protection
Agency, the Forest Service of the U.S. Department of Agriculture, and
the Fish and Wildlife Service and National Park Service of the U.S.
Department of the Interior.

Library of Congress Catalog Card Number 81-84469
International Standard Book Number 0-309-03196-6

Available from:

NATIONAL ACADEMY PRESS
2101 Constitution Ave., N.W.
Washington, D.C. 20418

Printed in the United States of America

iii

iv

ACKNOWLEDGMENTS

The Committee gratefully acknowledges the leadership of Ellis B. Cowling in initiating this study and in chairing the planning session that developed its scope. Thanks are also due to Barbara D. Brown and Raymond Herrmann of the National Park Service, Leon Dochinger and Donald Boelter of the U.S. Forest Service, Kent Schreiber of the U.S. Fish and Wildlife Service, and Dennis Tirpak of the U.S. Environmental Protection Agency. Their liaison role contributed richly to committee discussions and kept us informed of their agency's interests as they related to our study. We also thank G.J. Brunskill and J.H. Klaverkamp who reviewed the text and provided numerous references. The interaction with the NRC Report Review Committee was rewarding and beneficial to the report.

The Committee owes a special debt of gratitude to Robert Harriss, who contributed so generously of his time and thinking and helped in the writing of the report.

Finally, we thank the NRC staff members whose unflagging efforts and genuine interest made service on the Committee the enjoyable and successful experience that it was.

CONTENTS

FIGURES

TABLES

1. INTRODUCTION AND OVERVIEW

The hazards to humans and ecosystems from exposure to the high concentrations of atmospheric pollutants in urban areas are now well documented, having been the focus of much research. As a result, air quality considerations are becoming an integral part of urban and industrial planning.

On the other hand, when some of the same pollutants are disseminated broadly beyond urban areas, their fates and effects are still little known. The analytical methods sufficiently sensitive to measure trace concentrations of many pollutants have been developed only recently. But even our present rudimentary understanding of regional and global distribution of atmospheric pollutants, now coupled with improvements in the scope and sensitivity of toxicity testing, has produced a number of indications in the scientific literature suggesting that some pollutants, even when diluted by air masses on a continental or global scale, may accumulate in the biosphere to concentrations that may be proved to be harmful. Organochlorine pesticides and radionuclides from nuclear testing are among those cited for widespread dissemination and accumulation. More recently, evidence of other kinds of atmospheric pollution has been found in rural and even wilderness areas remote from major sources of atmospheric pollution. Examples are acid precipitation, increased trace metal deposition, and reduced visibility.

Many lines of evidence have shown that most kinds of atmospheric pollution have close associations with patterns of intensive human energy use. The consumption of fossil fuels is known to be the major source of anthropogenic pollution of the atmosphere, rivalled only by the high temperature smelting of metallic ores. It is predicted that energy demands will continue to increase exponentially (NRC 1979), and fossil fuels will play an important role in satisfying such demands. As a result, the Committee on the Atmosphere and the Biosphere decided to conduct a general review of the pollutants discharged to the atmosphere from fossil fuel burning.

Some of the effects of pollutants from fossil fuel usage have been relatively well studied--for example, the postulated linkage between increasing atmospheric CO_2 and climatic warming (NRC 1977a, Kellogg

1

and Schware 1981); the damage to vegetation by photochemical oxidants (NRC 1977b); and the health-related effects of SO_2, NO_x, and lead (NRC 1978a,d; 1980a). Therefore, this Committee focused its efforts on the broad scale ecological effects of SO_x and NO_x, as well as on the less well-known trace metal and organic pollutants. Attention was concentrated on the general properties of pollutants that govern their emission, dissemination, deposition, and ecological effects in the hope that discernible patterns would emerge that could assist management agencies and legislators in foreseeing some of the environmental consequences of continued fossil fuel burning, and in devising research programs that will address the most critical of unknown factors as quickly as possible.

THE FOSSIL FUEL SCENARIO:
THE PROBABILITY OF A CRISIS IN THE BIOSPHERE

Ecologists, geochemists, and climatologists are beginning to discover that in many respects man is now operating on nature's own scale, particularly through the heavy use of fossil fuels to supply the energy that runs our industrial civilization. Because the uncertainties associated with such large-scale operations are very great--for good or ill--it behooves us to exercise restraint in our present intensive use of energy, and to mitigate where possible the ill effects that air pollution imposes not only on us but also on the ecosystems that make up our life support system. In this connection numerous studies have reported on the substantial opportunities for energy conservation and shifts to alternative energy sources.

The predicted effects of the continued and accelerated combustion of fossil fuels are many and complicated. Some climatologists hypothesize that discharge to the atmosphere of CO_2 and other pollutants is likely to change the radiation balance of the earth sufficiently within the next century or so to cause major warming as well as changes in the patterns of climate. While there is considerable uncertainty in predictions, some of the hypothesized consequences could be severe. Droughts may be caused that are capable of reducing food production in many areas now providing surpluses for export. Other areas of the world may become able to produce more food than before, but many of these currently lack the technology and capital to do so. The postulated changes in climate would also cause major changes in many of the earth's natural ecosystems. Moreover, the hypothesized rise in sea level caused by the melting or breakup of the Greenland and west Antarctic ice sheets might necessitate the eventual relocation of major cities (NRC 1977a, Kellogg and Schware 1981), although the time scale is uncertain.

Perhaps the first well-demonstrated widespead effect of burning fossil fuel is the destruction of soft-water ecosystems by "acid rain," which has been caused by anthropogenic emissions of sulfur and nitrogen oxides that are further oxidized in the atmosphere. Major effects include destruction of many species of fish and their prey and

acidification of surface and ground waters to the point where toxic
trace metals reach concentrations undesirable for human consumption
and for aquatic animal habitats (see Chapter 8). Severe degradation
of many aquatic ecosystems has been recorded in Norway, Sweden, the
United States, Scotland, and Canada (Drablös and Tollan 1980).
Surprisingly, significant effects have also been recorded for aquatic
systems in countries once thought to be geologically resistant to
acidification, including the Netherlands, Denmark, and Belgium
(Vangenechten and Vanderborght 1980, Van Dam et al. 1980, Rebsdorf
1980).

Although claims have been made that direct evidence linking
power-plant emissions to the production of acid rain is inconclusive
(Poundstone 1980, Curtis undated), we find the circumstantial evidence
for their role overwhelming. Many thousands of lakes have already
been affected, according to estimates by European and North American
scientists (Drablös and Tollan 1980). Acidified lakes have been found
in areas where geological factors, such as volcanism and the
weathering of pyrites, or biological factors, such as acid bog
drainage, cannot be implicated. At current rates of emission of
sulfur and nitrogen oxides, the number of affected lakes can be
expected to more than double by 1990, and to include larger and deeper
lakes (Henriksen 1980). There is little probability that some factor
other than emissions of sulfur and nitrogen oxides is responsible for
acid rain. Although the deposited nitrogen and sulfur may be slightly
beneficial in terrestrial soils deficient in these elements, the
stimulus is expected to be short-lived (Abrahamson 1980), and over the
long term acid precipitation is likely to accelerate natural processes
of soil leaching that lead to impoverishment in plant nutrients
(Overrein et al. 1980). When freshwater effects are considered, the
positive effects are greatly outweighed by the negative.

The same gaseous oxides are known to affect human health directly
in urban areas (Lave and Seskin 1977, NRC 1977b, NRC 1977c, NRC 1978a,
Mendelsohn and Orcutt 1979) and to cause decreases in crop yields in
some areas. Acid deposition is also known to cause large economic
losses by corroding metals and eroding buildings and statuary made of
calcareous rock (Bolin 1971, Nriagu 1978). The Committee believes
that continued emissions of sulfur and nitrogen oxides at current or
accelerated rates, in the face of clear evidence of serious hazard to
human health and to the biosphere, will be extremely risky from a
long-term economic standpoint as well as from the standpoint of
biosphere protection.

Less well appreciated is the widespread dispersal of toxic metals
by the burning of fossil fuel (Bertine and Goldberg 1971, NRC 1980a).
Of most immediate concern are mercury, lead, zinc, and cadmium.
Mercury metal occurs at high concentrations in many fishes under
natural conditions. Its release to the atmosphere through the burning
of coal and crude oil, as well as through smelting, cement
manufacture, and municipal incineration, may already have been
sufficient to elevate mercury concentrations enough to make fish from
many areas unacceptable for human consumption. Lakes and rivers

containing fish with elevated mercury concentrations have been reported from Sweden, Quebec, New Brunswick, Ontario, Minnesota, New York State, and Maine (Tomlinson et al. 1980). Precipitation in areas receiving acid deposition and high sulfate deposition also has elevated mercury concentrations (Tomlinson et al. 1980, Semb 1978, Dickson 1980), and mercury uptake by fish appears to be enhanced by low pH. At present, there is no satisfactory technology for controlling large-scale emissions of mercury to the atmosphere. Its continued or accelerated release, especially in view of its synergism with acid deposition, may cause chronic problems in many areas in years to come.

Natural lead releases now have been exceeded by emissions from man's activities (Lantzy and Mackenzie 1979, NRC 1980a). While data are scarce, some have argued that the present body burden of lead in average Americans, and probably in other residents of industrial societies, is approaching chronically damaging levels (Patterson, see minority report in NRC 1980a; Settle and Patterson 1980). Leaded gasolines are heavily implicated, though other industries are also important sources of lead (NRC 1980a).

Models based on concentrations in sediment and water and on projected rates of increases in emissions suggest that in Lake Michigan both cadmium and zinc will reach concentrations toxic to zooplankton within the next 30 to 80 years depending on the rate of increase in emissions of the metals (Muhlbaier and Tisue 1981; Marshall and Mellinger 1980; G.T. Tisue and J. Marshall, Argonne National Laboratory, Argonne, IL, personal communication; Figure 1.1). Many other waters may be similarly affected, although careful studies have not been done yet to document these changes.

The chronicle of detrimental substances from the burning of fossil fuel could be continued for many pages, treating--among others --vanadium, arsenic, copper, and selenium. Also emitted are organic micropollutants of many types, some of which are known to be carcinogens, acute toxicants, teratogens, and mutagens. Hence, faced with the total array of atmospheric pollutants that could be reviewed--far beyond the scope of one committee's abilities--the Committee on the Atmosphere and the Biosphere decided to focus its attention on pollutants generated in energy production, most notably on sulfur and nitrogen compounds including acid rain, trace metals, and organic substances. Carbon dioxide, an important pollutant, is treated in some detail by NRC (1979b), and little space is devoted to it here.

THIS REPORT

Patterns of emission, transport, deposition, interaction, and biological effects of energy-related pollutants are known only incompletely, and we were forced to use a mosaic of examples to illustrate different concepts. We hope that this review will provide a preliminary guide to the sorts of integrated research needed to overcome weaknesses in our understanding of the consequences of

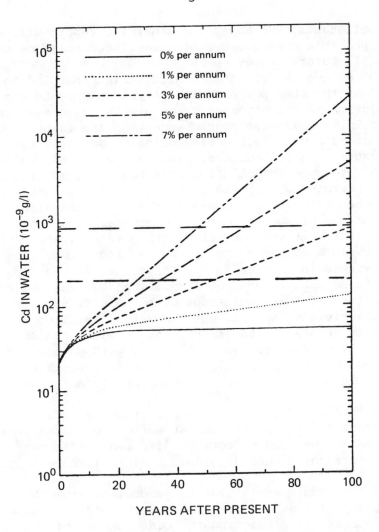

FIGURE 1.1 Projected cadmium concentrations in Lake Michigan for various rates of increase in the annual input rate. The lower horizontal dashed line represents the concentration at which toxic effects have been reported for aquatic organisms (Marshall and Mellinger 1980). The upper dashed line is the current U.S. EPA standard for waters of Lake Michigan's hardness. The present rate of increase of the input rate for Lake Michigan is 4.6 percent per annum, which would cause the lake to reach the toxicity threshold for aquatic organisms in about 40 years. The authors indicate that a high proportion of the input of cadmium is from the atmosphere. Cadmium in sediment at 10^{-6} g/g is equal to 1.5×10^{-8} g/l in water. SOURCE: After Muhlbaier and Tisue (1981). Copyright © 1981 by D. Reidel Publishing Co., Dordrecht, Holland.

atmospheric pollutants. We hope, too, that the report will make apparent the probable consequences of unregulated reliance on fossil fuels to fulfill future energy needs.

Chapter 2 reviews the scientific discoveries that led to the realization that the atmosphere is inseparably linked to the biosphere and that pollution of the atmosphere--even with trace amounts of toxicants--could have serious consequences for the latter. It is noteworthy that although atmospheric pollution has only recently become a widespread public concern, clearcut scientific evidence for the effects of atmospheric pollution from fossil fuel burning and smelting was generated during the second half of the 19th century.

Confusion has recently developed over natural emissions of substances from the biosphere to the atmosphere. The mechanisms of exchange of sulfur, nitrogen, reactive trace metals, and organic substances between atmosphere and biosphere, insofar as they are known, are reviewed in Chapter 3. In Chapter 4, we quantify the emissions of these substances from fossil fuel burning and compare the magnitude and form of these emissions with those from other important anthropogenic activities as well as from natural sources.

Chapter 5 is devoted to the transport, chemical transformation, and deposition of energy-related pollutants, with particular emphasis on physical and chemical properties of the emissions that affect their residence time in the atmosphere and thus the distance they are carried.

Chapter 6 is devoted to processes affecting pollutants once they enter the biosphere, with particular attention to similarities in pathways, repositories, and effects of different pollutants. Examples of synergism, antagonism, and development of resistance to pollutants are given.

Chapter 7 represents an attempt to generalize from what is known about important pollutants and sensitive parts of the biosphere to develop a blueprint for predicting the consequences of continued or accelerated pollution of the atmosphere. Again, because the information base is incomplete, only a rough, conceptual outline has resulted. Two of the major impediments to making more specific conclusions or predictions are the lack of a long-term data base sufficient to quantify small increases in toxicants and the inadequacy of methods for detecting and predicting low-level toxic effects, at the level of either organisms or ecosystems.

In Chapter 8, we give an up-to-date case history of acid deposition, one example of the severe biospheric effects resulting from anthropogenic pollution of the atmosphere that appears already to have eliminated or substantially reduced the populations of some organisms, including fishes, in parts of their natural ranges. Owing to the concentrated efforts of scientists in the Northern Hemisphere, most notably in Scandinavia during the past decade, we have a much more complete knowledge of the causes and consequences of acid deposition than we have for other pollutants discussed here. Yet even for acid deposition, many mechanisms and effects are poorly known. Throughout the above-mentioned chapters we have attempted to indicate how widespread and complicated the atmospheric pollution problem is, and how intertwined the biogeochemical cycles of pollutants are with the cycles of naturally important elements. As a consequence, second- and third-order effects may be of great importance, but they are far more difficult to predict than the first-order effects upon which most attention is focused.

Though necessarily incomplete in many respects, the information synthesized by the Committee renders a rather unfavorable picture of the consequences of current fossil fuel burning practices. Along with the probable large-scale effects described above, slow, nearly undetectable increases in a multitude of extremely toxic substances are taking place. It is the Committee's opinion, based on the evidence we have examined, that the picture is disturbing enough to merit prompt tightening of restrictions on atmospheric emissions from fossil fuels and other large sources such as metal smelters and cement manufacture. Strong measures are necessary if we are to prevent further degradation of natural ecosystems, which together support life on this planet.

Some pollutants, such as sulfur and nitrogen oxides, particulate pollutants, and lead, are readily amenable to control by appropriate engineering technologies. Others, such as mercury and CO_2, are more difficult and expensive to control by available technologies. In the long run, only decreased reliance on fossil fuel or improved control of a wide spectrum of pollutants can reduce the risk that our descendants will suffer food shortages, impaired health, and a damaged environment.

CONCLUSIONS AND RECOMMENDATIONS

Atmospheric pollution and its consequences deserve major consideration when the sources and sites of energy production are decided. However, much remains to be done if we are to assess adequately the ecological significance of atmospheric pollutants generated by different energy systems.

To improve and refine our ability to detect what may be irreversible degradation in natural ecosystems, increased and improved scientific effort is needed in two critical areas: long-term monitoring and forecasting of future effects, and ecotoxicology.

At present, the long-term data base on emission, deposition, and biological effects of all energy-related pollutants is insufficient. Such information is critical to forecasting ecological effects, both of fossil fuel combustion and of alternative energy production systems. Natural sources of the substances and linkages to the cycles of ecologically important chemicals must also be known, in order to discern how natural biogeochemical processes may be stressed or disrupted.

Likewise, our knowledge of the toxicity of pollutants requires rapid development in four major areas: (1) identification of the physico-chemical and biological properties of natural ecosystems and organisms that are sensitive indicators of stress from pollutants; (2) development of statistical and mathematical methods for detecting and quantifying ecological deviations significantly outside the normal range in unstressed ecosystems, particularly in response to the synergistic and antagonistic effects of several pollutants acting in concert; (3) identification and protection of ecosystems, communities, and species that are especially sensitive to present or projected pollution burdens; and (4) an evaluation of how current toxicity tests, usually done with single species, relate to stresses at the ecosystem level.

The personnel available to undertake such studies are not sufficient in number, nor are they trained adequately in the several disciplines necessary for the detection of subtle increases in pollutant stress on ecosystems over long periods. Protection of our threatened, sensitive ecosystems requires educational programs that are specifically designed to produce scientists with the necessary qualifications. While this need and some appropriate topics for study are identified and discussed in this report and elsewhere (NRC 1975, NRC 1981, Butler 1978), educational programs and research techniques need to be more clearly defined. In particular, current educational programs in toxicology must be upgraded to include rigorous statistical training and thorough grounding in ecological principles, with emphasis on detection of long-term, whole-ecosystem effects rather than on the current, short-term tests of individual species.

The needed curriculum should utilize the combined strengths of programs in ecology and environmental engineering. University programs of biology and ecology usually lack strength in applied aspects of ecology, while programs in environmental engineering often pay insufficient attention to the foundations of their discipline in ecology. While combined programs do exist at some institutions, many lack the scientific rigor necessary for training experts to attack the complex problems that will be faced in the coming years.

Current institutional arrangements are not designed to undertake effectively the complex, long-term studies envisaged. One possibility that deserves consideration is to set up a National Center for Ecology and Environmental Science, with functions similar in part to those of the National Center for Atmospheric Research and the U.S. Geological Survey. These organizations provide alternative models of governance--the first being supported by federal funds but governed by a consortium of universities, the second, a larger organization both funded and controlled federally. Such an institution would also require the cooperation and coordination of federal agencies, such as the U.S. Departments of Agriculture, Energy, the Interior, Health and Human Services, and the Environmental Protection Agency.

2. SCIENTIFIC UNDERSTANDING OF ATMOSPHERE-BIOSPHERE INTERACTIONS: A HISTORICAL OVERVIEW

Scientific appreciation of the linkage between the atmosphere and the biosphere has developed along several lines over the past four centuries. The first clue must have come from the act of breathing, and the work of Boyle, Hooke, Lower, and Mayow in the 17th century demonstrated that air contains an active component that is required both for breathing and burning (Dampier 1948, Wilson 1960, Taylor 1963). A century later, in 1780, Lavoisier and Laplace showed that as oxygen is consumed in breathing, carbon and hydrogen are oxidized into carbon dioxide and water that are given off along with heat (Gabriel and Fogel 1955).

That plants draw nourishment from the air around them, as well as from the water in the soil, was suggested by Hooke, on the basis of experiments by Thomas Brotherton in the 17th century (Hooke 1687, see also Gorham 1965). Other experiments by Hales (1738) supported this conclusion; the significance of Hooke's and Hales's observations could not be fully appreciated at the time, however, because the nature of atmospheric gases was not understood until the work of Scheele, Priestley, and Lavoisier in the latter part of the 18th century. It was Priestley (1772) who showed that plants can restore the capacity of air to support burning and breathing. Moreover, he recognized that this mechanism must be important in maintaining the capacity of the atmosphere to support animal life and combustion, thus intimating the interactions of the carbon and oxygen cycles in the atmosphere that maintain the balance of nature. Soon after, in 1779, Ingenhousz proved it was the green parts of plants that were necessary for

This chapter was prepared by Eville Gorham. It focuses strongly on acid rain, partly because of the current significance of the problem in relation to fossil fuel combustion, partly because the history of the problem has not been treated adequately elsewhere, and partly because of the author's own involvement with the subject. The author wishes to thank Leonard G. Wilson, Rudolph B. Husar, and Thomas C. Hutchinson for suggesting certain source materials.

restoration of the air, which took place only in sunlight (Gabriel and Fogel 1955). Between 1783 and 1788 Senebier showed that the chemical change involved conversion of "fixed air" (carbon dioxide) into "dephlogisticated air" (oxygen) and carbon, and in 1804 de Saussure made much more quantitative studies of the assimilation of carbon dioxide and water accompanied by the release of oxygen during photosynthesis (Nash 1957).

Both de Saussure and Senebier believed that the oxygen so released came from the carbon dioxide assimilated. It was not until the second quarter of this century that studies of photosynthetic bacteria by van Niel, of isolated chloroplasts by Hill, and the use of oxygen isotopes by Ruben and his associates (Gabriel and Fogel 1955) proved conclusively that oxygen came instead from water molecules, as suggested long before by Berthollet (Baker and Allen 1971).

The importance of the carbon dioxide cycled through the atmosphere and the biosphere to the process of rock weathering became appreciated in the first half of the 19th century--the major cycles of the atmosphere, biosphere, and lithosphere being thus tied together (Davy 1821, Jamieson 1856).

The involvement of the biosphere in the nitrogen cycle was demonstrated by de Saussure, whose water-culture experiments showed the stimulation of plant growth by nitrate (Baker and Allen 1971). Gay Lussac and Thenard recognized nitrogen as an important component of plants (Davy 1821), occurring in the form of gluten and also as albumen, more characteristic of animals. Liebig persuaded scientists by 1840 that plants could be produced from carbon dioxide, water, ammonia (NH_3), and certain minerals--the ultimate products of the decay of plants (Singer 1959, Holmes 1973). He thus controverted the older agricultural theories that plants utilize complex, soluble organic compounds released during the breakdown of manures (e.g., Woodward 1699, Hales 1738, Davy 1821). Liebig later found ammonia in plant sap, and came to believe that as it was also present in the atmosphere it must be absorbed by plants aerially like carbon dioxide in amounts sufficient to meet their full needs for nitrogen (Eriksson 1952).

Boussingault proved during the 1850s that most plants obtain their nitrogen from nitrate in the soil (Singer 1959). His earlier experiments, in the 1830s, had shown that legumes were an exception to this rule. He had suggested atmospheric fixation (Aulie 1973), but he could not explain the process, and a fuller understanding awaited the discovery in 1886, by Hellriegal and Wilfarth, that legumes reduce atmospheric nitrogen by means of nodules on their roots and can grow independently of other fixed nitrogen (Russell and Russell 1950). Biejernick cultured the nitrogen-fixing bacteria from nodules in 1888 (Collard 1976); Berthelot, already in 1885, had suggested that bacteria were responsible for the fixation he observed in soils (Crosland 1973).

The general involvement of soil microorganisms in the biological cycles of the elements was postulated in 1872 by Cohn to account for the breakdown of dead plant material and for the fact that organic manures restored the fertility of soils exhausted by continued cropping (Geison 1971, Collard 1976). Liebig had held that decay, putrefaction, and fermentation were purely chemical, but the role of microbes in these processes was becoming known through the work of Pasteur and others from 1857 on (Collard 1976, see also Smith 1872). (Pasteur's work on this subject was anticipated in 1837 by Schwann and in 1838 by Cagniard-Latour, but their results unfortunately were not accepted by Liebig and others; see Brock 1961.) The person who did most to confirm Cohn's hypothesis was Winogradski, who isolated several microbes involved in the cycles of nitrogen, sulfur, and carbon and investigated their metabolism (Collard 1976). By the 1890s the existence of heterotrophs, chemoautotrophs, and photoautotrophs was established, together with their ability to use and transform diverse organic or inorganic molecules or both in their energy metabolism. But only recently has much attention been devoted to the circulation through the atmosphere and the biosphere of a variety of gaseous compounds--such as carbon monoxide, methane, dimethyl sulfide, hydrogen sulfide, sulfur dioxide, ammonia, nitrous oxide, nitric oxide, and nitrogen dioxide--which contain biologically significant elements (cf. Hutchinson 1954, Goldberg 1974). Among the more interesting recent discoveries is the role of microbial methylation in the cycle of mercury and other elements. Although the vapor-phase transfer of elemental mercury from the soil to the atmosphere is of major importance, the formation of monomethyl mercury, a potent neurotoxin, probably enhances such transfer, locally. Dimethyl mercury is even more volatile (U.S. Geological Survey 1970, Wood 1974.)

An aspect of atmosphere/biosphere interactions that was discovered long ago is the involvement of plants in returning water to the atmosphere. In the mid-17th century van Helmont observed that a 169-pound willow tree had grown in 5 years from a 5-pound stem planted in a tub of soil and watered only with rain or distilled water (Baker and Allen 1971). He believed the tree's substance to be transmuted water. Woodward (1699) refined van Helmont's experiment and concluded that most of the water a plant used was evaporated via pores to the atmosphere, especially during warm weather, and that the plant grew from the terrestrial vegetable and mineral material contained in all waters, even rain.

Early in the following century Hales (1738) made many careful measurements of the absorption of water and its transpiration to the atmosphere, in one case comparing the amount of water transpired from the plants in a hop field with the amount evaporated from the soil. At the same time that the above-mentioned studies were going on, Halley was examining the nature of the hydrological cycle described by Aristotle and quoted by William Harvey (1628), and he was demonstrating quantitatively that evaporation from the ocean and from watercourses is adequate to replenish river flow (Biswas 1970a, b). The hydrologic role of plant transpiration was recognized by de la Methiere in 1797, according to Biswas (1970b).

The mineral nutrition of plants was first suggested, as noted above, by Woodward. In 1733 Tull, a celebrated agriculturist, reiterated the view that minute earthy particles of soil were the true nourishment of plants (Davy 1821). It was left to de Saussure to demonstrate by analysis in the early 19th century the uptake of diverse mineral elements from the soil (Taylor 1963). A table of his data was given by Davy (1821), who indicated that alkalis and alkaline earths were essential ingredients of plants and that silica might serve as a structural material.

The possibility that certain plants might depend upon atmospheric sources for some of their mineral nutrients was suggested by several authors as long ago as the 17th century (Guerlac 1954), and again in the 18th century by Hales (1738). Early in the following century a few agriculturists (Naismith 1807, Aiton 1811, Dau 1823) realized that the plants of certain kinds of peatlands were not influenced at all by water that has percolated through mineral soil but depended solely upon rain and snow (see DuRietz 1949; Gorham 1953, 1978a). Smith (1852) placed the earlier conjectures upon a more scientific basis when he pointed out that rainwater contains enough nutrients (such as ammonia) to allow plants to grow, though scantily. The chemical composition of surface waters from rain-fed peat bogs was first analyzed by Ramann in 1895 (Kivinen 1935). Witting (1947, 1948) contrasted the ionic composition of rain-fed bog waters with that of fen waters receiving water from the mineral soil, and Gorham later (1961) demonstrated the chemical similarity of bog waters and rain waters. The geochemical contrast between bog and fen peats was shown by Mattson, Sandberg, and Terning (1944) and Mattson and Koutler-Andersson (1954). Recently, study of the atmospheric inputs to the nutrient budgets of ecosystems has greatly increased (e.g., Galloway and Cowling 1978).

The early history of research on the elements and ions in atmospheric precipitation dealt largely with the study of nitrogenous compounds, as pointed out in a thorough review by Eriksson (1952). Nitrate was observed in rainwater by Marggraf in the winter of 1749-50, along with calcium and chloride (see Miller 1913, who reviewed the earliest literature of rain chemistry). Ammonia was found in the atmosphere by Scheele in 1788-89 and by de Saussure in the early 1800s. In the latter half of the 19th century, numerous investigations of atmospheric nitrogen supply were made, chiefly in Europe, but also in Asia, Africa, and New Zealand. Among them, those of Boussingault in the 1850s, cited by Smith (1872), are of particular interest. Boussingault showed a clear decline of ammonia concentration during the course of a given rainfall and also observed especially high concentrations in fog and dew.

Systematic studies of chloride were also made in the mid-19th century, when the role of sea spray was clearly demonstrated by Smith (1872), who set up a network of stations in the British Isles to make a short-term study of rain chemistry. He also pointed out the abundance and importance of sulfate in urban areas (Smith 1852, 1872). The general nature of the cycles of chloride and sulfur was given a major review by Eriksson (1959, 1960).

Only in the last few decades have thorough analyses been made of the full suite of major cations and anions in atmospheric precipitation. Networks for long-term regional studies are equally recent; the first was the Scandinavian (later European) network

(Emanuelsson, Eriksson, and Egnér 1954). Data on trace elements and specific organic constituents (both major and minor) in air and precipitation are mostly recent and are still scattered and sporadic (see, e.g., Katz 1961; Stocks, Commins, and Aubrey 1961; Lazrus, Lorange, and Lodge 1970; Lunde et al. 1976; Galloway and Cowling 1978). In the 19th century, however, some trace elements were observed in rain, such as iodine (Pierre, quoted by Smith 1872) and arsenic (Russell, quoted by Cohen and Ruston 1912).

Two general topics in the subject of atmosphere/biosphere interactions deserve mentioning. The first is the major evolutionary change about two billion years ago from a reducing to an oxidizing atmosphere, owing to the development of photosynthesis. This was apparently still a somewhat controversial matter even after World War II (Hutchinson 1954), although it had been proposed by Goldschmidt in 1933 (Rankama and Sahama 1950). The evidence from paleobiology and geochemistry has been thoroughly reviewed by Cloud (1968, 1976). The second topic is the reciprocal fitness of the evolving atmosphere and biosphere, in agreement with the classic argument of Henderson (1913) that "fitness of environment is quite as essential as the fitness which arises in the process of organic evolution."

HUMAN ALTERATIONS OF THE ATMOSPHERE

Radioactivity in rainfall was observed as long ago as 1906 (Eriksson 1952), but it has received serious attention only in the era since World War II, as a result of concern over the threat to human health posed by radioactive fallout from atomic weapons. Early work in the 1950s was thoroughly reviewed by Caldecott and Snyder (1960). The many radioisotopes that were released proved to have an unanticipated utility in tracing both physical and biological aspects of atmosphere/biosphere interactions (Nelson and Evans 1969, Broecker 1974). Among the most interesting phenomena was the striking concentration of several isotopes along the food chain from mosses and lichens to reindeer and caribou and thence to Laplanders and Eskimos (Gorham 1958a, 1959; Lidén 1961; Miettinen 1969).

The buildup of carbon dioxide in the global atmosphere due to the combustion of fossil fuels has also become a major concern, because of the possibility that it may increase the temperature of the atmosphere and alter the world's climatic patterns quite substantially through the so-called greenhouse effect. The development of our knowledge of this problem has been reviewed by Plass (1956). He reports that Fourier in 1827 compared the atmosphere to a pane of glass beneath which the earth is warmed. The role of carbon dioxide in the greenhouse effect was mentioned by Tyndall in 1861 and worked out by Arrhenius in 1896. Over the next three years Chamberlin presented in detail the geological implications of the carbon dioxide theory of climatic change.

Callendar (1938) suggested that combustion of fossil fuel by man was enriching the atmosphere in carbon dioxide sufficiently to induce perceptible climatic warming. He later (Callendar 1949, cf.

Hutchinson 1954) suggested that land clearance and cultivation might also be important. The magnitude of the carbon dioxide release was shown clearly by Keeling (1970).

If continued, the carbon dioxide enrichment and consequent climatic warming could have diverse and serious consequences for the biosphere, including major displacements of agriculture (Kellogg 1978; Committee on Government Affairs 1979; see, however, Idso 1980). Such displacement would come about because of the predominant role of climate in determining the distribution of the biota. The influence of climate has of course been known since antiquity, but early in the 19th century it was studied scientifically by the great geographer von Humboldt (e.g., von Humboldt 1805). The modern view of an interactive control of plant distribution by temperature, moisture, topography, and geology was clearly expressed shortly thereafter by Watson (1833, cf. Gorham 1954).

Depletion of the stratospheric ozone layer could result from microbial transformation of nitrogenous fertilizers (Crutzen 1970), emissions of water vapor and nitrogen oxides from supersonic transports (Harrison 1970, Crutzen 1970, Johnston 1971), and the use of chlorofluorocarbons (freons) in refrigerators and especially as propellants in aerosol sprays (Molina and Rowland 1974, Wilkniss et al. 1975). Serious damage to the ozone layer would greatly increase the penetration of ultraviolet radiation through the atmosphere, with a variety of damaging effects upon plants and animals--including an increase in human skin cancers (NRC 1976a).

Chlorinated hydrocarbons, such as the pesticide DDT and the industrial PCBs, everywhere undergo a distillation to the atmosphere that allows them to spread throughout the world ecosystem (Goldberg 1975). DDT is likewise spread throughout the atmosphere on particles of talc used as a carrier and diluent in aerial pesticide sprays and is observed ubiquitously in atmospheric dust samples (Windom, Griffin, and Goldberg 1967). These toxicants are of especial concern because they can accumulate significantly along food chains. Such biological magnification was discovered in the 1950s, and the early work is cited by Rachel Carson (1962) in "Silent Spring."

The injection of fine particulate pollutants, including a variety of sulfates and condensed hydrocarbons, is interfering with atmospheric visibility on a regional scale (NRC 1979a). The possibility of atmospheric particulates forming from chemical reactions of gases was pointed out long ago by Rafinesque (1819, 1820). The role of air pollution in reducing local visibility has probably been known ever since coal came into use as a major energy source and was noted by Evelyn in 1661 (Brimblecombe 1977). Crowther and Ruston (1911) and Cohen and Ruston (1912) remarked on the substantial reduction in hours of sunshine and light intensity by air pollution in the city of Leeds. More recently, Flowers, McCormick, and Kurfis (1969) showed that atmospheric turbidity is unusually high in the heavily polluted eastern United States, and there has been a distinct trend over the past half century toward increasing turbidity (McCormick and Ludwig 1967) both in and near cities. Increasing regional haze in the eastern United States and its possible climatological consequences were discussed by Husar and his colleagues (1979). The nature of photochemical smog of the Los Angeles type was first elucidated by Haagen-Smit (1952, 1953). Its effects were first investigated by Middleton and his associates (1950, 1958, 1961), and have been reviewed recently by the National Research Council (1977b).

Atmospheric haze may also be a consequence of natural biospheric activity such as the large-scale production of terpenes by coniferous and other kinds of vegetation. This effect has received rather little investigation since its identification by Went (1960; see also NRC 1976b; Curtin, King, and Mosier 1974; Simoneit and Mazurek 1980). Natural eolian transport of partially biogenic dust also contributes to haze. Its transport over long distances from Africa to the Atlantic Ocean was described in the mid-19th century by Darwin (1846) and Ehrenberg (1849, cf. Simoneit 1979).

Acid rain is one of the most serious and far-reaching results of air pollution. The phenomenon was noted by Hales in 1738; he remarked that dew and rain "contain salt, sulphur, etc. For the air is full of acid and sulphureous particles . . .," and he said that these constituents "make land fruitful" after plowing. (Nash, 1957, pointed out that these particles were not to be taken as the materials themselves, but as their alchemical principles). The true local significance of acid rain appears to have been recognized first by Smith (1852), who from his analyses of rain in and around the heavily polluted city of Manchester, in England, remarked, "We may therefore find easily three kinds of air, -- that with carbonate of ammonia in the fields at a distance, -- that with sulphate of ammonia in the suburbs, -- and that with sulphuric acid, or acid sulphate, in the town" (Smith's italics). In the same article Smith pointed out that free sulfuric acid in city air was responsible for the fading of colors in prints and dyed goods and the rusting of metals.

Twenty years later Smith (1872), then General Inspector of Alkali Works for the British government and a Fellow of the Royal Society, produced a classic account of the chemistry of air and rain, comparing country sites in England, Scotland, and Ireland with heavily populated urban sites in England, Scotland, and Germany. In this account he noted several important patterns (see Figure 2.1):

- the decline of rain chlorides away from the seacoast, with the exception that city rains are secondarily enriched by chlorides from coal combustion. (Smith also remarked that even in coastal rains the ratio of sulfate to chloride was greater than in sea water).
- the presence of abundant sulfuric acid in urban rain, particularly in industrial areas because of the combustion of coal rich in sulfur. (Smith was the first to use the term "acid rain.")
- the liberation of hydrochloric acid into urban atmospheres by the interaction of sulfuric acid and sodium chloride during or after coal combustion (cf. Gorham 1958c,d; Eriksson 1958; Odén 1964; Hitchcock, Spiller, and Wilson 1980).

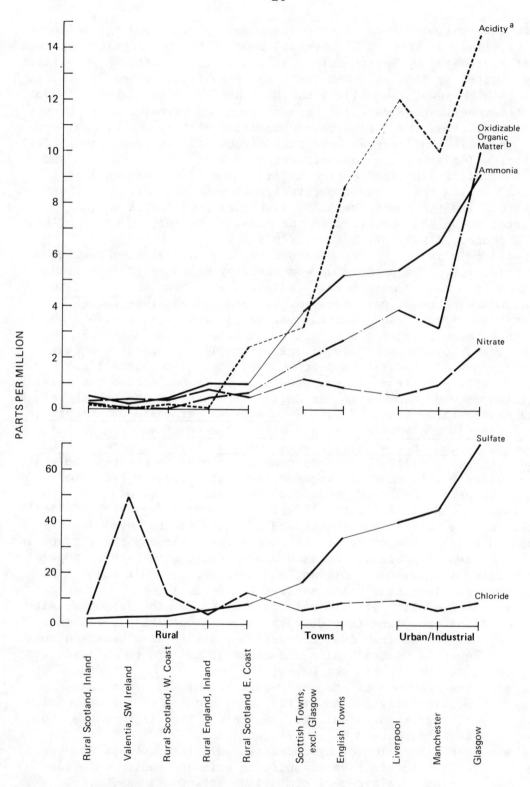

FIGURE 2.1 Chemical composition of rain in the British Isles reported in 1872.
SOURCE: Data from Smith (1872).

- the liberation of sulfur and ammonia into the air, both by decomposition of dead organic matter in country areas and by coal combustion in cities, and their deposition together in rain.
- the likelihood of arsenic, copper, and other metals occurring in urban and industrial atmospheres.
- the general lack of vegetation in cities where the air contains enough acid to yield rains with 40 ppm of sulfuric acid.
- the interference of acid gases with development of the grain in wheat.
- the bleaching of chlorophyll in aquatic plants by very dilute acids.
- the possibility that acid gases from factories might weaken plants and expose them to attacks by fungi less susceptible to the same gases.
- the damage to stone, brick, mortar, iron, galvanized iron, and brass by acid rain.

Smith quoted extensively from studies in 1854 and 1855 by a Belgian commission on damage to plants by acid emanations from chemical industries. The commission pointed out that such damage was related to numerous environmental factors such as distance from source, temperature, humidity, rainfall, wind direction and frequency, path of the smoke plume, topography, and shelter by obstacles to wind currents. Chimney height was noted as important. The commission described--and produced experimentally--a variety of types of damage to plants, including leaf spots and bleaching of chlorophyll, marginal leaf damage, early leaf-fall, and damage to buds and young twigs. It observed, moreover, considerable difference in sensitivity to acid gases, both between species and within varieties of the same species. In the course of its work it tested the acidity of raindrops upon plant surfaces with blue litmus paper and found the raindrops to be acid only near the chemical works whose effects they were examining.

Further work upon acid rain in and near urban areas was done by Crowther and Ruston (1911; see also Cohen and Ruston 1912, Crowther and Steuart 1913) in and near the city of Leeds, England. They reported levels of total suspended matter, ash, tar, soot, nitrogen, sulfur, chloride, and acidity declining away from the center of the city. Cohen and Ruston (1912) also compared wet and dry fallout and the scavenging of suspended solids by rain and by snow. Crowther and Ruston (1911) showed that air pollution inhibited the growth of plants and their power to assimilate carbon dioxide. By watering soils with Leeds rain and with similarly dilute sulfuric acid, they arrested seed germination and growth and inhibited three aspects of the nitrogen cycle--ammonification, nitrification, and nitrogen fixation. Timothy grass under such regimes became distinctly poorer in protein and richer in crude fiber. Cohen and Ruston (1912) demonstrated the clogging of leaf stomata by soot, the narrowing of tree rings owing to emissions from newly constructed shale works, and several changes in soil quality (including loss of carbonates) after leaching by acid rain and by similar concentrations of sulfuric acid.

The high acidity of Leeds rain was ascribed to sulfuric acid, although Crowther and Ruston noted that coal combustion greatly increased both sulfate and chloride levels in the center of the urban area. In fact, if one does a partial correlation of acidity at Crowther and Ruston's 11 stations with both anions, to eliminate the influence of correlation between the two anions, it appears that only the correlation ($\underline{r}$) with chloride is significant:

$\underline{r}$ values (n = 11)	zero order*	significance	first order**	significance
H^+ on Cl^-	0.795	$p < 0.01$	0.703	$p < 0.05$
H^+ on $SO_4^=$	0.543	$p < 0.05$	0.176	not significant

* Influence of anion intercorrelation included.
** Influence of anion intercorrelation eliminated.

Gorham (1958c), examining data compiled by Parker (1955), found a similar situation in two other cities in northern England. It appears that hydrochloric acid from coals rich in chlorine predominates in urban precipitation there, whereas sulfuric acid from the oxidation of drifting sulfur dioxide (slowed in the urban area by the presence of hydrochloric acid) predominates in rural rain (Gorham 1955, 1958d). Another early student of acid rain, Bottini (1939), found hydrochloric acid abundant in precipitation near the volcano Vesuvius.

Damage to vegetation by sulfur dioxide from metal smelters has a long history. According to Almer and his colleagues (1978), it was recorded in 1734 by Linnaeus, on a visit to the 500-year-old smelter at Falun in the Swedish province of Dalarna. Smelter problems--including also livestock poisoning by arsenic--were investigated extensively in the United States both by observation and by experimental fumigation early in the 20th century (Swain 1949). A long report by Holmes, Franklin, and Gould (1915) includes an extensive annotated bibliography of early American and German studies of the effects of sulfur dioxide upon plants and animals, including humans. Effects of sulfuric acid--resulting from fumigations with sulfur dioxide--upon the calcium status of soils poor in lime were also reviewed. Katz and his associates (1939) reported local acidification of soils and a lowering of their base saturation by sulfur dioxide emissions from a lead-zinc smelter in British Columbia. Severe local damage to vegetation by sulfur dioxide fumigation was also described by Katz, who reviewed the work in this field back to Schroeder and Reuss in 1883.

A number of other investigators between 1939 and 1954 observed strongly acid to alkaline pH values in precipitation. Landsberg (1954) cited five beside himself, to whom may be added Atkins (1947) and Tamm (1953). Tamm also measured pH in rain passing through forest canopies and the concentrations of several other elements. A long series of unpublished pH measurements since about 1930, at three adjacent stations a little to the north of London, England, has also been discovered very recently (Brimblecombe and Pitman 1980). It reveals annual average pH varying around 4.6 between 1930 and about 1965-70, declining to about 4.0 today.

The year 1955 saw a new emphasis on the study of acidity in precipitation, and the recognition of its spread from urban and industrial sources to distant rural areas, with the analysis of extensive data sets from the international network in Scandinavia by Barrett and Brodin (1955), from a number of British cities by Parker (1955), and from the rural English Lake District by Gorham (1955). Houghton (1955) examined numerous samples of fog and cloud water from New England. Gorham continued to study the chemistry of atmospheric precipitation, with a series of papers on its nature, origin, and influence upon the geochemistry of oligotrophic lake waters, bog waters, and soils, culminating in a general review of the subject in 1961. He also examined, in a series of four papers with Gordon, the effects of fumigation with sulfur dioxide--and resultant acid rain--upon terrestrial and aquatic ecosystems around metal smelters in Ontario (see Gordon and Gorham 1963). The Scandinavian (later European) network has provided the basis for a series of papers by diverse authors.

International transboundary air pollution was noted long ago by Evelyn (1661), who remarked that farmers in parts of France to the southwest of England complained of smoke driven from England's coasts, which injured their vines in flower. By the mid-18th century, London's urban smoke plume was sometimes observable at distances of 100 km (Brimblecombe 1978). Major international concern about acid rain as a serious, widespread pollution problem with severe ecological consequences began with the publication (Bolin 1971) of Sweden's report to the United Nations Conference on the Human Environment, which arose out of studies by Odén (1967, 1968). Attention was drawn to the problem in the United States by Likens and his associates, beginning early in the 1970s (Likens, Bormann, and Johnson 1972; see also Likens 1976). These authors noted the likely significance in the United States of nitric acid produced by the further oxidation of nitrogen oxides emitted from gasoline engines. The predominance of nitric acid in rain at Pasadena, California, has been reported recently by Liljestrand and Morgan (1978). An historical resumé of progress in scientific and public understanding of acid precipitation and its biological consequences has been given by Cowling (1981).

ENERGY AND AIR POLLUTION

Acid rain is generally the result of severe air pollution by man's combustion of fossil fuels (for exceptions see Bottini 1939 and Hutchinson et al. 1979a). The record of pollution by coal smoke and of government action about it goes back at least to the 13th century in England, and the problem appears to have become extremely serious by the end of the 17th century (Shaw and Owens 1925, Brimblecombe

1975, 1976, 1977, Brimblecombe and Ogden 1977, Halliday 1961, Heidorn 1978, Lodge 1980). That air pollution from coal combustion is a serious factor in human mortality, especially from lung disease, was postulated as long ago as the mid-17th century by the Londoner John Evelyn (1661) in his "Fumifugium." He also noted corrosion of structures and materials and losses of plants and animals, and remarked that in 1644--when Newcastle (the source of "Sea Coale") was blockaded--gardens and orchards produced far more than in the years before and after.

Evelyn suggested that industries, rather than residences, were the chief cause and called for their banishment to five or six miles downriver (usually downwind) from the city. The editor of the 1772 edition of Evelyn's "Fumifugium" claimed a marked worsening in the situation over a century, and called for a variety of ameliorative measures: the use of tall smokestacks to spread the smoke into "distant parts," better chimney construction to drive the smoke higher, changed methods of combustion using charred (coked) coal to lessen smoke emission, inducements for industries to move outside the city, and legal prevention of more such building within the city. Tall smokestacks had been suggested as early as the late 14th century (Lodge 1980), and coal cleaning was apparently tried as long ago as the 15th century (Yeager 1979). Other early suggestions, such as a shift to wood or to anthracite from bituminous coal, seem not to have been followed (Brimblecombe 1975).

The deleterious effects of air pollution (smoke) upon health and mortality were noted by John Graunt (1662) the founder of the science of statistics and demography. He reported that London was becoming increasingly unhealthful and blamed this largely upon the great increase in the use of coal over the preceding sixty years, although he did believe that population growth and crowding were partly responsible. Among the first detailed studies of the influence of coal smoke upon human health (as measured by mortality from nontubercular lung disease) were those of Ascher, given in an appendix by Cohen and Ruston (1912).

Lichens appear to be the plants most sensitive to air pollution, and Grindon in 1859 attributed the decline of certain species in the city of Manchester, England to pollution (Hawksworth and Seaward 1977). According to Skye (1968), the absence of these plants from urban areas was recorded in Paris by Nylander in 1866. Further details about the influence of air pollution upon lichens are given by Barkman (1958) and by Ferry, Baddeley, and Hawksworth (1973).

Twentieth-century studies of the effects of air pollution upon human health, vegetation, and corrosion have been reviewed by the World Health Organization (1961) and by Higgins and his associates (NRC 1979a). Health effects have been examined statistically in great detail by Lave and Seskin (1977) and Mendelsohn and Orcutt (1979). Their work has been questioned by Lipfert (1980) and McCarroll (1980). Modern legislation against air pollution, beginning in the 19th century, has been reviewed by Halliday (1961) and Lodge (1980), who also described the evolution of control technologies.

The development from local, point-source pollution of the kind seen in London or Los Angeles to the broad dispersion of acid rain over large parts of the European and North American continents has come to be recognized as a major anthropogenic perturbation of atmosphere/biosphere interactions.

3. BIOGENIC EMISSIONS TO THE ATMOSPHERE

A wide variety of natural products, including volatile substances and particulates, are emitted into and removed from the atmosphere by both terrestrial and aquatic systems. Terrestrial reactions are better known, owing to the large body of agricultural research, and knowledge about the compounds thus generated or removed is increasing constantly. Assessments will need to be modified frequently as information is obtained on factors affecting rates of transformation and on local, regional, and global emissions and removals.

Soils emit a variety of volatile products and also remove volatile substances from the gas phase. The activities of microorganisms dominate the soil ecosystem, and most evidence indicates that heterotrophic microorganisms are the chief agents for the production of many gases. The biochemical mechanisms involved in the microbial generation of these gases vary markedly, however, and no generalization covers all of the volatile products or even any significant class of volatiles.

NITROGEN

Terrestrial Ecosystems

Ecosystems consume atmospheric molecular nitrogen (N_2) through biological nitrogen fixation and also evolve it as one of the products of denitrification. Biological fixation of N_2 leads to incorporation of nitrogen compounds into soil. The annual global fixation of molecular nitrogen by the land mass has been estimated by a number of authors. One such estimate indicates that the annual amount of molecular nitrogen fixed into soil globally is 99×10^{12} grams of nitrogen per year (Delwiche and Likens 1977).

Denitrification is the process by which some microorganisms can reduce nitrate (NO_3^-) or nitrite (NO_2^-) to the gaseous forms molecular nitrogen (N_2) and nitrous oxide (N_2O), which may then be lost to the atmosphere. This is a major pathway for the loss of fixed nitrogen from an ecosystem. An enormous amount of work is being done on denitrification, and the topic has been reviewed recently (Delwiche and Bryan 1976, Delwiche et al. 1978).

Some estimates of denitrification have been made by measuring the quantity of nitrous oxide generated. Most of these studies have been conducted in the laboratory, but a number of studies under field conditions (Focht and Stolzy 1978, Denmead 1979) have shown that nitrous oxide generation may occur in both dry-land agriculture and in flooded fields (Denmead et al. 1979). The rate of denitrification varies extensively with season of the year and oxygen content of the soil (Dowdell and Smith 1974). The major factors that affect nitrous oxide emissions are nitrate content of the soil, oxygen status, moisture, pH, and temperature.

It is generally believed that nitrous oxide is formed by the microbial reduction of nitrate or nitrite as microorganisms use these compounds for the terminal electron acceptor in their metabolism. An intriguing recent investigation, however, suggests that some nitrous oxide may also be generated in soils during nitrification, the microbial oxidation of ammonia to nitrite and nitrate (Bremner and Blackmer 1978, Freney et al. 1978). This formation of nitrous oxide from ammonia was earlier shown to take place in cultures of autotrophic nitrifying bacteria (Yoshida and Alexander 1970).

Nitric oxide (NO), too, is also produced in soil. Presumably, this gas is generated in denitrification as nitrate is reduced. Nitrate is the dominant if not the sole precursor of nitric oxide. The quantity of nitric oxide formed is markedly affected by temperature, pH, and moisture (Bailey and Beauchamp 1973, Garcia 1976), and much of the release is apparently microbial (Garcia 1976).

Calculations have been made of the total quantity of inorganic nitrogen volatilized through denitrification. One estimate is 120×10^{12} grams of nitrogen per year (Delwiche and Likens 1977). An early estimate of the emission of nitrogen oxides gave values for biological release globally of 120×10^{12} grams nitrogen per year of nitrous oxide and a figure of 234×10^{12} grams nitrogen per year of nitric oxide (Robinson and Robbins 1970a). A more recent estimate suggests that the quantity of nitrogen oxides emitted from natural sources on the earth's surface in the Northern Hemisphere is probably no more than 30×10^{12} grams of nitrogen per year (Galbally 1976). It is obvious from the range of these estimates that the data base is limited, and the conceptual models are crude.

In making such measurements, it is difficult to separate evolution and uptake of nitrogenous gases. Nitrous oxide is both emitted and removed from the gas phase. There is some evidence that net removal takes place largely in waterlogged soils (Freney et al. 1978). Denitrification is promoted by anaerobiosis and by the presence of organic materials that stimulate microbial proliferation. Altogether, soils are a more significant source than a sink for nitrous oxide (Blackmer and Bremner 1976).

Agricultural scientists and technologists have extensively studied ammonia losses from soil. Most of these studies reflect anthropogenic inputs of nitrogen because they have been done on fertilized, tilled soils or pastured areas. Many studies were prompted by concern with the loss of the nitrogen applied to the soil in the form of fertilizer for crop production. In some instances, losses can be quite appreciable; for example, in one study, it was noted that the ammonia-nitrogen loss varied from 19 to 50 percent of the quantity of fertilizer nitrogen applied, the extent of loss being dependent on the fertilization rate and the temperature (Fenn and Kissel 1974).

Despite this large body of information, the inputs of ammonia to the atmosphere from large land areas have not been measured directly except in a few instances. In one such study, the flux of ammonia over a grazed pasture in Australia was determined for a 3-week period in late summer. The quantity of ammonia evolved was equivalent to 260 grams per hectare per day (10.8 g/ha/hr), a quantity that is a substantial part of the nitrogen cycled in the pasture. It was estimated that, under these conditions, a major source of the ammonia was probably the urine of the sheep in the pasture (Denmead et al. 1974). Other studies have indicated that the losses in Australian pastures may be up to 13 grams of nitrogen per hectare per hour when the pasture is grazed but only about 2 grams of nitrogen per hectare per hour when it is not grazed (Denmead et al. 1976).

A number of investigations have verified the significance of animal manure, particularly from dairy cattle, to the volatilization of ammonia and its presence in the overlying air (Luebs et al. 1973, Lauer et al. 1976). Data from the U.S. National Atmospheric Deposition Program (Gibson and Baker 1979) show large amounts of ammonia in precipitation in areas of the midwest where feedlots are abundant and where anhydrous ammonia is a popular agricultural fertilizer.

Global estimates, based upon unfortunately few data, have been made by several individuals. The consensus has been that microbial action in soil is a major source of ammonia generated from the land surface but the size of estimates varies greatly. One estimate of the ammonia loss from the soil is 960×10^{12} grams per year (Robinson and Robbins 1970a); another investigator has suggested that the natural emissions of ammonia in the Northern Hemisphere do not exceed 130×10^{12} grams of ammonia nitrogen per year (Galbally 1976).

Several organic nitrogen compounds, including amines and nitrogen heterocycles, may also be evolved from soils. Research has focused attention on the generation of these products in lands treated with animal manure, either from cattle or poultry. Among the amines formed are mono-, di- and trimethylamine, ethylamine, other alkylamines, indole, and skatole. These products are probably the result of microbial activity in decomposing animal wastes, and the process is probably also occurring under the anaerobic conditions that prevail in heaps of these organic materials (Young et al. 1971, White et al. 1971, Rouston et al. 1977, Mosier et al. 1973). Volatile organic nitrogen, presumably as amines, may be detected not only immediately above the decaying organic matter but also in traps placed at sites adjacent to high densities of animals (Elliott et al. 1971).

Aquatic Ecosystems

Much information on nitrogen in aquatic ecosystems is summarized in a recent National Researach Council report (1978d). We shall therefore review only new or especially relevant aspects here. Nitrogen fixation in aquatic ecosystems is carried out largely by a few genera of blue-green algae (Brezonik 1977, Jones and Steward 1969). These genera become abundant in lakes where the supply of nitrogen is low relative to that of phosphorus (Schindler 1977). A survey of recent literature revealed that atmospheric molecular nitrogen is not fixed in lakes that have inputs of nitrogen 10 times greater by weight than phosphorus (Flett et al. 1980). Nor did this survey find fixation of N_2 in either oxic or anoxic lake sediments, although nitrogen was fixed by the periphyton community of a small lake with a very low input ratio of N to P. In two small lakes, one studied for three years, nitrogen fixation supplied 7 to 38 percent of the total annual income of nitrogen. Rates of nitrogen fixation depended on light intensity, much as carbon fixation does.

It is impossible to make large-scale estimates of N_2 fixation in fresh water, because there are few quantitative studies for large water bodies. Howard et al. (1970) showed that algal fixation occurred in Lake Erie but supplied no quantitative estimates. Vanderhoef et al. (1974) showed that fixation supplied only a small part of the nitrogen input to Green Bay, a highly eutrophic part of the Laurentian Great Lakes. This makes it seem unlikely that most other large lakes, which are more oligotrophic, fix substantial amounts of N_2, and lakes are probably minor participants in the global nitrogen cycle.

A technical problem renders uncertain many quantitative estimates of nitrogen fixation in either terrestrial or aquatic systems. Most studies to date have employed the convenient, sensitive, and inexpensive acetylene-reduction method, which assumes that every three moles of acetylene reduced to ethylene by the nitrogenase system represent the equivalent of one mole of molecular nitrogen fixed (Graham et al. 1980). In two comparisons, however, the actual conversion ratio has varied between 1.7 and 9.1 moles of acetylene reduced per mole of nitrogen fixed (Hardy et al. 1968, Graham et al. 1980). Because of this variability in the conversion ratio estimates of total nitrogen fixed based solely on the acetylene method may be in error by a factor of two to four.

Denitrification rates are highest in eutrophic lakes, where the combination of a rich organic substrate, high nitrate and low oxygen favors the denitrifying microbiota (Brezonik 1977). The epilimnion sediment-water interface appears to be the most important site (Tirén et al. 1976, Chan and Campbell 1980), although the thermocline region can also be important when sufficient nitrate is present along with low oxygen (< 0.2 mg/l). Denitrification can also take place in anoxic hypolimnia, with nitrous oxide often occurring as an end product. The importance of denitrification at this site is limited, because high nitrate concentrations usually do not occur under anoxic conditions.

Chan and Campbell (1980) estimate that only 1.4 percent of the nitrate entering a small eutrophic Canadian lake was denitrified. They found that a combination of nitrate-nitrogen (>0.01 mg/l) and oxygen (<0.2 mg/l) was necessary for significant denitrification to take place.

In coastal and marine sediments there is some evidence that ammonia may be a significant end product of nitrate reduction (Koike and Hattori 1978, Sorensen 1978). Regional data for aquatic ecosystems are too few to warrant global estimates.

SULFUR

Terrestrial Ecosystems

It is generally accepted that soils are sources of volatile sulfur compounds. Estimates have given widely dissimilar figures; emissions from the land surface, where the reaction is generally believed to be biological, for example, have been estimated at 68×10^{12} (Robinson and Robbins 1970b), 58×10^{12} (Friend 1973), and as little as 3×10^{12} grams of sulfur per year (Bolin and Charlson 1976). The authors of most of the estimates believe that the sulfur emitted biologically from soil is derived from vegetation decaying in intertidal flats, swamps, and bogs, and it has generally been assumed that the product evolved was hydrogen sulfide. But monitoring studies and direct analysis of soils indicate that any hydrogen sulfide formed would probably react with iron compounds to form iron sulfide (Bloomfield 1969).

Volatile sulfur compounds are formed in soil samples treated with sulfur-containing amino acids. Such soils evolve methyl mercaptan, dimethyl sulfide, dimethyl disulfide, ethyl mercaptan, ethyl methyl sulfide, diethyl disulfide, carbon disulfide, and small amounts of carbonyl sulfide. Some of these products are detected during the decomposition of animal manure (Banwart and Bremner 1975), and soils treated with a variety of natural materials, including sewage sludge and plant remains, evolve some of the same gases, but no hydrogen sulfide is emitted. The conversions are affected by the dominant amino acids present and by the level of oxygen (Banwart and Bremner 1976a, b). Hence, the suggestion that hydrogen sulfide emissions are major sources of the sulfur coming from soil is apparently incorrect. No large areas of the globe that evolve hydrogen sulfide have been discovered. The chief product coming from adequately aerated soil is more likely to be dimethyl sulfide (Rasmussen 1974). There are only limited measurements of the release of volatile biogenic sulfur compounds from soil (Adams and Farwell 1980, Aneja et al. 1980, A. Goldberg et al. 1981), and therefore the global significance of soil as a source is difficult to evaluate at this time.

Aquatic Ecosystems

The reduction of sulfate to sulfide occurs in fresh waters, marine areas, and fresh and salt marshes under anoxic conditions. The bacterial genus Desulfovibrio is responsible, and rates are apparently highest in tidal salt marshes, where tides replenish the supply of sulfates. In such salt marshes, sulfate reducers actually contribute as much energy to the salt marsh food chain as direct photosynthetic fixation. Howarth and Teal (1979) report an annual rate of 75 mol SO_4 per square meter for a New England marsh. Highest rates occurred in autumn, when decaying plants increased the available organic substrate.

In fresh waters, production appears to be largely in anoxic hypolimnia or in sediments rich in organic matter. Measurements are too few to allow reliable regional or global estimates. The rate of reduction is dependent on sulfate concentrations, as well as on the available organic matter, both of which vary widely in different lakes.

The amount of sulfide that actually reaches the atmosphere under natural conditions is not known. As in terrestrial soils, iron concentrations in the active areas are often high enough to exceed the solubility product of ferrous iron and sulfide, and in these cases the element is precipitated as iron monosulfide or pyrite (Howarth 1979, Schindler et al. 1980a, Cook 1981). In salt marshes, this precipitation is thought to prevent the buildup of high concentrations of soluble sulfide, which could be toxic to marsh grasses and other microbiota (Howarth and Teal 1979). Much of the iron sulfide is reoxidized as part of the seasonal cycle. Too few ecosystems have been investigated to permit assessment of how widespread the above phenomena may be. Both lakes and marshes may vary greatly in their iron content, depending on underlying geology. In iron-impoverished areas, substantial volatile sulfide may be released to the atmosphere.

Both marine and freshwater plants appear to produce dimethyl sulfide and dimethyl sulfoxide as metabolic by-products. When emitted to the atmosphere, these methylated sulfur compounds are rapidly oxidized to sulfur dioxide or sulfuric acid. The subject is reviewed by Andreae (1980). Lovelock et al. (1972) measured dimethylsulfide and carbonyl sulfide in air over the Atlantic Ocean, and assumed the compounds to be of marine origin. Subsequently, Nguyen et al. (1978) measured dimethylsulfide in sea water and calculated that the oceans could contribute substantially to the natural global sulfur budget. Maroulis and Brandy (1977), however, estimated dimethylsulfide emissions from Atlantic coastal regions of the United States to be less than 6 milligrams per square meter per year, which would make an inconsequential contribution to the natural global sulfur budget. Thus, total natural sulfur emissions from aquatic ecosystems and wetlands cannot be quantified owing to the paucity of studies. Likewise, the small number and extreme variability of estimates of volcanic emissions of sulfur dioxide makes it extremely difficult to quantify total natural emissions of sulfur dioxide to the atmosphere. Available estimates of sulfur dioxide emissions for single volcanoes vary by a factor of 10 (Hoff and Gallant 1980).

TRACE METALS

There are several lines of evidence that terrestrial plants release metals to the atmosphere. For example, pea plants grown in solutions containing radioactive zinc released the metal to the atmosphere (Beauford et al. 1975). The rate of zinc mobilization was about 1 microgram/h/m^2 of leaf. Any extrapolation of these results to environmental situations is unwarranted; still, the data do suggest that there may be an appreciable flow of zinc from plants into the atmosphere.

Supporting evidence comes from field studies of plant exudates (Curtin et al. 1974). Polyethylene bags were placed around pine and fir trees, and the exudates, condensed in these bags over period of five to ten days, were collected and analyzed. The ash of the residue of volatile exudates contained lithium, beryllium, boron, sodium, magnesium, titanium, vanadium, chromium, manganese, iron, cobalt, nickel, copper, zinc, gallium, arsenic, strontium, yttrium, zirconium, molybdenum, silver, lead, bismuth, cadmium, tin, antimony, barium, and lanthanum. The underlined elements were most markedly enriched in the exudates compared to the ash of the associated vegetation. Curtin and colleagues attributed the volatilization of some of the metals to complexing with terpenes and urged the initiation of an air sampling program to assess this possibility.

Biological methylation on both land and sea may result in the transfer of metals from the earth's surface to the atmosphere. It is believed that mercury can enter the air from the sea surface in the form of gaseous dimethyl mercury (Wood and Goldberg 1977, Tomlinson et al. 1980). In the atmosphere it may disproportionate into ethane, methane, and mercury. Methylated forms of mercury are also produced in sediments, and diffusion into the overlying water may lead to these substances entering the atmosphere. Methylmercury formation is thought to be almost solely the result of microbiological processes (Woolson 1977). Various mercuric compounds can be used as substrates, including inorganic compounds, elemental metal, and acetate (Rogers 1979, Hamilton 1972).

Several investigators have demonstrated that mercury in soil can be converted to volatile products. The chief if not the sole agents of this volatilization are microorganisms, because the process is reduced markedly or abolished by sterilization of the soil (Rogers and McFarlane 1979). As in aquatic systems, the conversion of the element from a nonvolatile to a volatile state occurs with a number of forms of inorganic mercury as well as with mercuric acetate (Rogers 1979). One of the products, has been identified as methylmercury, the production of which is affected by soil moisture, temperature, and mercury concentration. High soil temperatures appear to be favorable to the production of the volatile derivative, but the conversion is retarded by excessive moisture or dry conditions (Rogers 1976, Landa 1979). Elemental mercury is also a significant source of mercury emissions from soil. Most of the mercury in the atmosphere appears to be in the elemental form (William Fitzgerald, Department of Chemistry, University of Connecticut, Storrs, personal communication). The dimethyl mercury emitted into the atmosphere later decomposes to elemental mercury.

These investigations with mercury have prompted studies with other trace metals that can be methylated and volatilized from plant surfaces or internal plant tissues, such as arsenic, selenium, tin, germanium, and lead (e.g., Beauford et al. 1975, Jernelov and Martin 1975). Concern has been expressed about the possible toxic effects of the lipid-soluble metalalkyls of the above metals upon the central nervous system of higher organisms. Methylated tin compounds have been measured in sea waters and algae from the coast of California. Dimethyl and tetramethyl tin compounds have been found in plants at concentrations usually exceeding those of inorganic tin (Hodge et al. 1979). In laboratory experiments, moreover, the protonated tin compound, stannane (SnH_4), is produced during the decomposition of plants. These tin compounds are volatile enough to be introduced from surface waters to the atmosphere where they are easily oxidized. It should be noted that human activities appear to have increased the tin levels in marine and fresh-water systems by about an order of magnitude (Seidel et al. 1980).

There is a large enrichment of selenium in the atmosphere, and Duce et al. (1974) have suggested that this selenium is evolved by natural rather than anthropogenic sources. By analogy to the microbial transformation of sulfur, which is similar to selenium, one might expect that the product would be a methyl selenium derivative. Indeed, studies of a number of individual microorganisms indicate that they can convert selenate, selenite, elemental selenium, and organic selenium compounds to dimethylselenide. One of the organisms studied can also produce hydrogen selenide (Doran and Alexander 1977a).

Studies under laboratory conditions indicate that selenium may be volatilized when soils are amended with elemental selenium, selenite, selenate, and some organic selenium compounds. These emissions apparently are the result largely or entirely of the activity of microorganisms, and the processes are promoted if organic materials are added to the soil. The chief product appears to be dimethyl-selenide, although dimethyl diselenide may be formed from certain organic sulfur compounds, and hydrogen selenide is possibly emitted as well (Doran and Alexander 1977b).

Microorganisms have been obtained in axenic cultures that can use certain volatile organic selenium compounds as carbon sources for growth (Doran and Alexander 1977a). The significance of this in vitro metabolism of organic selenium compounds to natural conditions and the significance of the products of these reactions to selenium levels in the atmosphere remain unknown. According to Lewis (1976), selenium is released from the leaves of selenium-accumulating Astragalus sp. (locoweeds) as organo-selenium compounds, including methyl- and dimethylselenides.

Inquiries into problems associated with human poisoning at the turn of the century determined that arsenic-containing wallpaper components were being converted by microorganisms into volatile arsenic derivatives (Gosio 1897). More recently, it has been established that microorganisms are able to convert a number of arsenic compounds, both inorganic and organic, to volatile alkyl arsenic derivatives (Cox and Alexander 1973).

Soils treated with arsenicals emit volatile products. In the early investigations of flooded soils, volatile arsenic products were not identified (Epps and Sturgis 1939). Lately, the products have been identified as methylarsenic derivatives, and the recent studies point to the possibility of soils thus transferring arsenic to the overlying atmosphere (Cheng and Focht 1979, Woolson 1977).

Figure 3.1 summarizes the biological cycles of two trace metals, mercury and arsenic. Microorganisms are known to metabolize Hg, As, Sn, Se, and S and form methylated compounds of these metals. A number of organisms can methylate tellurium (Fleming and Alexander 1972, Challenger 1951), and methylation reactions characterize the behavior of certain microorganisms when exposed to a number of toxic elements.

ATMOSPHERIC EMISSIONS FROM BURNING OF NATURAL BIOMASS

Supplementing such low-temperature biological processes is the burning--both natural and man-induced--of plants, which can introduce substantial quantities of metals as well as carbon, nitrogen, and sulfur compounds into the atmosphere. The exact amounts are not known, but forest fires often create temperatures of 400°C and up, clearly sufficient to vaporize some inorganic compounds. In the United States alone, forest fires have been estimated to contribute annually 27 megatons (2.4×10^{13}g) of particulates to the atmosphere and an additional 17 megatons (1.5×10^{13}g) per year are estimated to originate from slash burning and forest litter control operations (Robinson and Robbins 1971). There is in general a latitudinal zonation of forest types about the earth's surface. Tropical forests span the equatorial regions, but the "combustion" of their organic matter is usually through low-temperature, biologically mediated processes. The global contribution from forest fires of particles in the atmosphere is estimated to be 150 megatons (1.4×10^{14}g) per year.

Recent studies have raised interesting new questions concerning the influence that burning of biomass--wood and other natural products excluding fossil fuels--may have on global atmospheric budgets of certain trace gases such as CO_2, CO, H_2, N_2O, NO, NO_2, and COS (Adams et al. 1977; Wong 1978; Radke et al. 1978; Crutzen et al. 1979; Seiler and Crutzen 1980). These studies have used very limited, preliminary data on emission factors and total biomass burned to calculate regional and global inputs from burning. Estimates of atmospheric input of CO_2 from burning have been particularly controversial because of the extremely important climate implications. Field studies currently in progress may improve our understanding of emissions from large-scale intentional burning.

The growing popularity of wood as a fuel for domestic heating and certain industrial heat needs has also stimulated considerable effort to understand environmental effects on air quality (Cooper 1980; Hall and DeAngelis 1980). A summary of some measured emissions of major

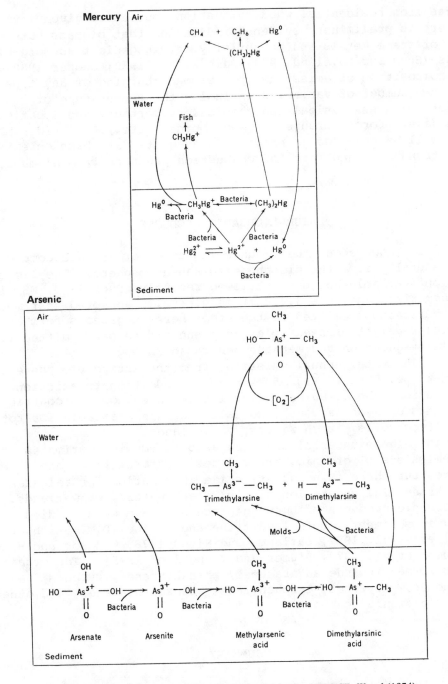

FIGURE 3.1 The biological cycle for mercury and arsenic. SOURCE: Wood (1974). Reprinted with permission from *Science* 183:1049-1052. Copyright © 1974 by the American Association for the Advancement of Science.

pollutants from residential wood combustion sources is given in Table 3.1. There is preliminary evidence indicating that biomass burning is a source of trace metals and potentially carcinogenic trace organic compounds (Shum and Loveland 1974; Bumb et al. 1980; Cooper 1980).

The composition of emissions from biomass burning of any type depends on a number of variables including the temperature of combustion, biomass physical and chemical properties, and location of the burn (i.e., forest, range, or wood burner, etc.). Considerable research will be required to provide an adequate data base on emission factors, transport, and fate for combustion products from biomass burning.

NATURAL ORGANIC PRODUCTS

Little is known about the amounts and types of organic compounds emitted naturally from the biosphere to the atmosphere. The limited information available has been reviewed recently (Duce 1978, NSF 1979) and is summarized in Table 3.2. It is generally accepted that most natural hydrocarbons emitted to the atmosphere in gaseous form are of low molecular weight--for example, isoprene and terpene emitted from terrestrial vegetation and methane generated in bog and lake sediments. The total annual emissions from the entire biosphere are estimated to be 850×10^{12} g of carbon. In addition to emissions from vegetation, the ocean and natural fires are known to contribute small amounts of hydrocarbons. No estimates are available for total emissions from soils, fresh waters, or wetlands.

Likewise, large uncertainties are associated with estimates of natural emissions of organic particulates. Natural fires are known to contribute, but there is no quantitative information for releases from oceans, terrestrial vegetation, soils, fresh waters, or wetlands.

The organic products of burning processes have been studied primarily in sediments, both lacustrine and marine (D.M. Smith et al. 1973, Müller et al. 1977, Laflamme and Hites 1978, Windsor and Hites 1979, Blumer et al. 1977, Blumer and Youngblood 1975). The organic compounds formed include anthracenes, phenanthrenes, pyrenes, fluoranthenes, chrysenes, triphenylenes, benzanthracenes, and others, as well as elemental carbon.

SUMMARY

A number of volatile products are generated from soils, vegetation or water surfaces both under natural conditions and after stimulation of the biota by anthropogenic activities. These volatile products include compounds of nitrogen, sulfur, numerous heavy metals, and organic substances. Both low-temperature, microbially mediated processes and high-temperature volatilization, as in forest fires, are responsible. The release of aerosols may be important in many cases. In all cases, data are too few to allow meaningful interpretation. Global estimates that have been made for emitted substances are little better than order-of-magnitude values.

TABLE 3.1 Emissions of Major Pollutants from Residential Wood Combustion

Chemical Species	Wood-Burning Stoves			Fireplaces		
	Grams per Kilogram of Wood	Pounds per 10^6 Btu	Percentage Particulates	Grams per Kilogram of Wood	Pounds per 10^6 Btu	Percentage Particulates
Carbon monoxide	160 (83–370)	22		22 (11–40)	3.0	
Volatile hydrocarbons	2.0 (0.3–3.0)	0.28		19	2.6	
NO_x as NO_2	0.5	0.07		1.8	0.25	
SO_x as SO_2	0.2	0.03				
Aldehydes	1.1	0.15		1.3	0.18	
Condensable organics	4.9 (2.2–14)	0.67	58	6.7 (5.4–9.1)	0.92	74
Particulates	3.6 (0.6–8.1)	0.50	42	2.4 (1.8–2.9)	0.33	26
Total particulates	8.5 (1–24)	1.2	100	9.1 (7.2–12)	1.3	100
Polycyclic organic material	0.3	0.04	3.5	0.03	0.004	0.3
Benzo(a)pyrene	0.0025	0.0003	0.03	0.00073	0.0001	0.008
Carcinogens[a]	0.038	0.005	0.45	0.0059	0.0008	0.06
Priority pollutants[b]	0.41	0.06	4.8	0.063	0.009	0.7
Na	0.005	0.0007	0.06	0.004	0.0006	0.04
Al	0.004	0.0006	0.05	0.002	0.0003	0.02
Si	0.003	0.0004	0.04	0.002	0.0003	0.02
S	0.03	0.004	0.4	0.004	0.0006	0.04
Cl	0.05	0.007	0.6	0.05	0.007	0.6
K	0.07	0.01	0.8	0.05	0.007	0.5
Ca	0.004	0.0006	0.05	0.005	0.0007	0.05
Organic carbon	4.2	0.58	49	4.2	0.58	46
Elemental carbon	0.7	0.1	8	1.2	0.16	13

[a]Includes benz(a)anthracene, dibenzanthracene, benzo(c)phenanthrene, benzofluoranthenes, methylcholanthene, benzopyrenes, dibenzopyrenes, and dibenzocarbonzoles.

[b]Includes acenaphthylene, fluorene, anthracene/phenanthrene, phenol, fluoranthene, pyrene, benz(a)anthracene, benzofluoranthenes, benzo(a)pyrene, benzo(ghi)perylene, dibenzanthracenes, acenaphthene, and ethyl benzene.

SOURCE: Cooper (1980). Cooper used values from a variety of sources discussed in his paper. Reprinted, with permission, from *Journal of the Air Pollution Control Association* 30:855–861.

TABLE 3.2 Natural Sources of Organic Carbon ($\times 10^{12}$ g of carbon per year)

Source	Gaseous Nonmethane Hydrocarbons	Particulate Organics
Vegetation		
Isoprene	350	?
Terpenes	480	
Others	?	
Total	⩾830	
Soil	?	11[a]
Ocean/freshwater	1.7	14
Biomass burning	3.5–46	1.6–7.1
Total	~830–880	~15–21[a]

[a]Soil-derived organics are mostly associated with large particles and are not included in the global budget.

SOURCE: Adapted from Duce (1978) and National Science Foundation (1979).

4. ANTHROPOGENIC SOURCES OF ATMOSPHERIC SUBSTANCES

An enormous number and variety of anthropogenic sources emit substances into the atmosphere. Manufactured products are atomized and vaporized; particulate matter and dust are released from construction, mining, and industrial activities; gases and vapors form at high temperatures during the combustion of fossil fuels, ore smelting, and cement manufacturing.

In this report we concentrate on the combustion of fossil fuels because this is the source of a major part of the anthropogenic substances in the atmosphere (see, e.g., Bertine and Goldberg 1971, Keeling and Bacastow 1977, Robinson 1977, Galloway and Whelpdale 1980, Shinn and Lynn 1979). Fossil fuels represent the largest mass of raw materials subject to high-temperature combustion processes. Certain types of coal and petroleum are enriched with potentially toxic metals (e.g., Hg, Se, As, Cd, and Zn) as well as radioactive elements and organic compounds, many of which are released to the atmosphere during combustion. Significant changes in the qualitative and quantitative patterns of fossil fuel consumption are expected during the next few decades with consequent changes in emissions to the atmosphere.

Some emissions from the combustion of fossil fuels enter the atmosphere as gases, such as sulfur dioxide, nitrogen oxides, elemental mercury, and volatile organic compounds (U.S. EPA 1978a, Morris et al. 1979, Lindberg 1980). Others enter as solid or liquid particles, the so-called primary aerosols (Robinson 1977, Block and Dams 1976). In addition, so-called secondary aerosols form from the gases--e.g., sulfur dioxide is transformed into ammonium sulfate (Husar et al. 1978, NRC 1978a). Volatile inorganic trace elements can also undergo transformations to particulate form during dispersion and cooling of combustion gases (Kaakinen et al. 1975). Atmospheric transformations, transport, and deposition of fossil fuel pollutants will be discussed in the next chapter.

PATTERNS OF FOSSIL FUEL USE

Global patterns of fossil fuel consumption and associated emissions have evolved over the past century in response to demographic, economic, and technological factors. Recent papers by H. Brown (1976), Häfele and Sassin (1977), and others have documented historical patterns of fossil fuel use. We review selected data on spatial and temporal patterns of fossil fuel use here to illustrate the potential for changing patterns of anthropogenic emissions to the atmosphere.

A central task for an assessment of atmosphere-biosphere interactions is the development of emissions inventories and projections. But because quantitative data on fuel consumption are generally available for most countries, while emissions data are often of poor quality or are not available at all, we must focus on consumption. With fuel consumption data, projections of emissions can be made using knowledge or assumptions concerning variables such as combustion technology, emissions-control technology, and patterns of operation (e.g., seasonal variations in electrical energy demand).

Total consumption of fossil fuel energy has grown at an average rate of 5 percent per year since 1900. Primary fossil fuel consumption over the past few decades is summarized in Figure 4.1. In the early stages of growth in fossil fuel use, coal was the most widely used fuel, whereas since 1950 oil and gas have become predominant, accounting for approximately 73 percent of present primary energy consumption (Häfele and Sassin 1977). The pattern of increasing global energy demand and fossil fuel use reflects the transition from a totally agricultural society to a partially industrialized one. Industrial development has been accelerated by the transition from a wood-fueled economy to a coal-fueled and then to an oil- and gas-fueled one; each successive energy source has provided increased energy efficiency with consequent positive feedback increasing total consumption. There has also been a tendency to centralize the location of fossil fuel burning usually near large urban areas.

Figure 4.2 shows proportional shares of the global energy market that each of these 4 fuel sources has supplied over time. Marchetti (1975) has demonstrated that primary fuel transitions to date have all exhibited similar dynamics. Each new energy source has taken roughly 100 years to grow from 1 percent to 50 percent of the global energy market. These data have important implications for the development of inventories and projections of global pollutant emission. An emissions forecast must include proper time dependence on the introduction of new pollution sources and the phasing out of replaced sources.

Most pollutants do not have sufficiently long residence times in the atmosphere to become globally distributed (see Chapter 5). With the exception of CO_2, lead, mercury, and perhaps a few other compounds, energy-related emissions are primarily of concern at present because of potential effects on atmospheric and ecological processes within a local, interregional, or continental area. Thus,

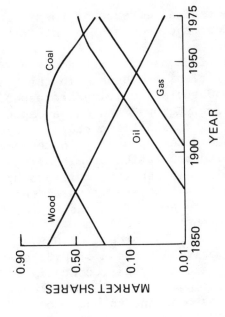

FIGURE 4.2 Global market shares of primary energy forms. SOURCE: After Häfele and Sassin (1977). Reproduced with permission from the *Annual Review of Energy*, vol. 2. Copyright © 1977 by Annual Reviews, Inc.

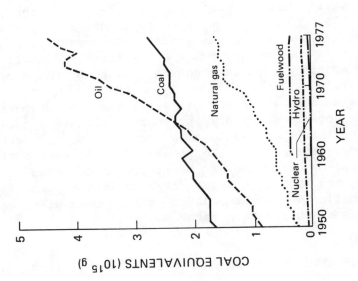

FIGURE 4.1 World energy production from primary sources, 1950-1977. SOURCE: After Sivard (1979). Copyright © 1979 by World Priorities.

an assessment of potential effects of fossil fuel use on environmental quality must consider the geographical distribution of fossil fuel emissions.

Regional energy flux density is shown in Figure 4.3. About 90 percent of fossil fuels are consumed in the Northern Hemisphere. Three centers of industrial activity are especially important: eastern North America, Europe, and China-Japan region. As we shall discuss in Chapter 5, this has important implications for pollutants that are transported interregionally in the hemisphere but not globally. A detailed tabulation of annual energy consumption for each nation can be obtained in reports published by the United Nations (1976, 1978). The past increase in total energy consumption and shifts from one fuel to another in the United States have largely paralleled the patterns of global use.

Coal represents over 90 percent of U.S. fossil fuel reserves but currently supplies only 20 percent of U.S. energy needs. Its use will probably grow as a result of the recent oil crisis and public opposition to nuclear power. While total coal consumption has remained more or less constant through most of this century (Figure 4.4), there has been a dramatic shift in the economic sectors that consume coal. Before 1940, coal consumption was divided among railroad power, residential and commercial heating, oven coke production, and other industrial processes. The railroad demand was particularly high during the war years of the mid-1940s. Then, within one decade, the 1950s, coal consumption by railroads and by the residential-commercial sector all but vanished. Currently, electric utilities are the main coal consumers, and the trend of total coal use in the United States since 1960 has been determined by the demands of the electric utilities.

The result has been that, in recent years, a higher proportion of coal emissions has come from large point sources with very tall smoke stacks. This means that the emissions can spread over a larger area before they fall or are washed to the ground. At the same time, increased use of air-conditioning has meant increased demand for electricity during the summer, a season that may favor more rapid chemical transformation in the atmosphere of the power plant emissions because of the higher ambient temperatures.

ATMOSPHERIC EMISSIONS FROM FOSSIL FUEL BURNING

In the absence of direct measurements of emissions, an assessment of atmospheric emissions of specific substances is dependent on quantitative inventories of the total mass (e.g., of coal used) or surface exposure (e.g., for volatile emissions from natural soils or vegetation) of the various sources and their chemical composition. An emission flux is obtained by multiplying the mass of the source by the appropriate emission rate. Some data on rates of fossil fuel use and raw-material refining for production of industrial materials are available in publications by the United Nations, World Bank, U.S. Department of Energy, and U.S. Bureau of Mines. Emission rates,

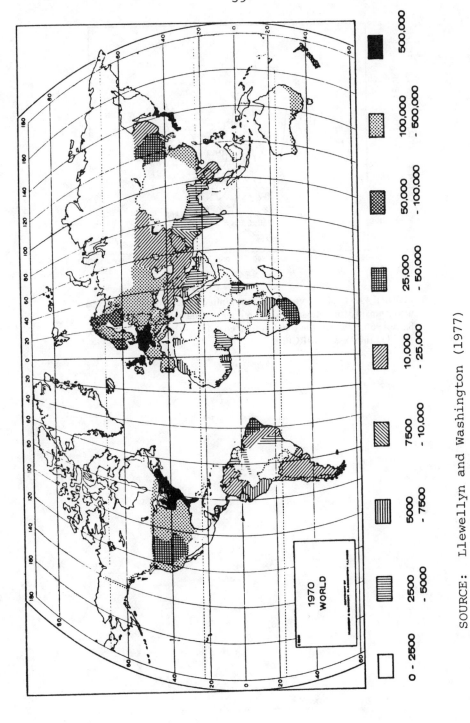

| | 0 - 2500 | | 2500 - 5000 | | 5000 - 7500 | | 7500 - 10,000 | | 10,000 - 25,000 | | 25,000 - 50,000 | | 50,000 - 100,000 | | 100,000 - 500,000 | | 500,000 |

SOURCE: Llewellyn and Washington (1977)

FIGURE 4.3   World energy flux density expressed as kilograms of coal per square kilometer per year, 1970. SOURCE: Llewellyn and Washington (1977).

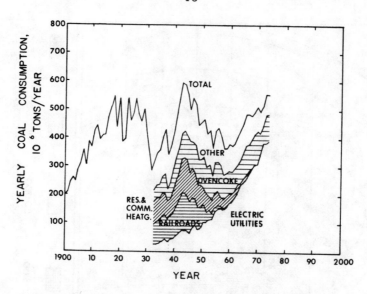

FIGURE 4.4 Coal consumption in the United States. Initially, coal was used primarily for railroads and for residential and commercial heating. Since 1960, however, the trend in total coal use has been determined by electric utility use. SOURCE: NRC (1978a).

particularly for high priority, health-related pollutants, are compiled by the U.S. Environmental Protection Agency (U.S. EPA 1977).

A comparison of natural and anthropogenic emissions at the global, national, and regional levels is a first key step in a careful assessment of potential ecological consequences. An example of this approach for sulfur sources in eastern North America is summarized in Table 4.1.

Oxides of Sulfur and Nitrogen

Sulfur dioxide was emitted from man-made sources in the United States at an estimated rate of 30×10^{12} g per year in 1973 (Figure 4.5). Fuel combustion exclusive of transportation accounted for 78 percent, industrial processes (metal smelting, chemical industries and manufacturing, etc.) for 20 percent, and transportation for 2 percent. With the exception of fuels used in transportation, about 65 percent of the national anthropogenic sulfur oxide emissions came from coal combustion and about 13 percent from oil combustion.

Some 85 percent of the sulfur dioxide emitted in the United States is released east of the Rocky Mountains, with the highest emission density in the vicinity of the Ohio River Valley (Ohio, Pennsylvania, and Indiana). The emission density for the states in the Ohio River Valley region ranges between 10 and 30 g/m^2 per year of sulfur dioxide. East of the Rocky Mountains coal contributes 71 percent and oil, 20 percent of the total sulfur dioxide emissions. Hence, coal combustion and, to a lesser degree, oil consumption are a proper index of the emissions of sulfur oxides (SO_x) over the eastern United States, and state-by-state trends of coal consumption are useful indicators of regional SO_x emission trends.

Emissions from electric utilities constitute a growing share of the total SO_x emissions (Figure 4.5). In 1973, utilities contributed about 60 percent of the SO_x emissions east of the Mississippi. According to EPA estimates, SO_x emissions will increase by about 30 percent in Texas by 1985. Even with more stringent standards applied to new power plants SO_x emissions are predicted to continue at about their present level while the nitrogen oxide (NO_x) emissions are projected to increase (Figure 4.6).

The national NO_x emissions of about 22×10^{12} g per year arise in roughly equal proportions from automobiles, industry, and electric utilities. The emission density of NO_x, as seen in Figure 4.7, is highest in the Boston-Washington corridor, and second highest in the Ohio River Valley states of Indiana, Ohio, and Pennsylvania. The oxides of nitrogen play important roles in a wide range of atmospheric processes which include the formation of aerosols, photochemical reactions in both the troposphere and stratosphere, the formation of acid precipitation, and the degradation of air quality in urban areas (NRC 1977c and 1978d, Crutzen 1979, Kelly et al. 1980). Figure 4.8 illustrates the major sources, pathways, and removal mechanisms for atmospheric nitrogen species—exclusive of molecular nitrogen—that are of particular significance to atmosphere-biosphere interactions.

TABLE 4.1   Atmospheric Sulfur Budget for Eastern North America ($\times 10^{12}$ g/year)

	Magnitude for Eastern		
	Canada	U.S.A.	North America
Inputs			
Man-made emissions	2.1	14	16
Natural emissions,			
sea spray, internal	0.06	—	0.06
terrestrial biogenic	0.06	0.04	0.1
marine biogenic	0.2	0.4	0.6
Inflow from oceans	0.04	0.02	0.06
Inflow from west	0.1	0.4	0.5
Inflow to U.S. from Canada	—	0.7	—
Inflow to Canada from U.S.	2.0	—	—
Total	4.6	15.6	17.4
Outputs			
Wet deposition	3.0	2.5	5.5
Dry deposition	1.2	3.3	4.5
Outflow to oceans	0.4	3.9	4.3
Outflow from Canada to U.S.	0.7	—	—
Outflow from U.S. to Canada	—	2.0	—
Total	5.3	11.7	14.3

SOURCE: Galloway and Whelpdale. Reprinted with permission from *Atmospheric Environment*, vol. 14, J.N. Galloway and D.M. Whelpdale, "An Atmospheric Sulfur Budget for Eastern North America". Copyright ©1980 by Pergamon Press, Ltd.

43

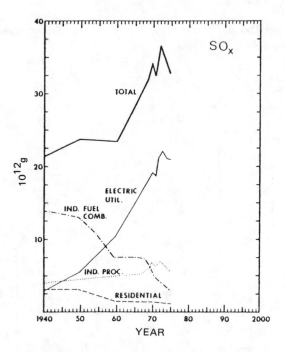

FIGURE 4.5   Sulfur oxide emissions in the United
States by source, 1940-1975. From 1940 to 1960 the
reduction of SO_x emissions from industrial fuel consumption was balanced by increase of SO_x emissions
from electric utilities. Since 1960 the sharp increase
in SO_x emissions was essentially due to electric
utilities. SOURCE: NRC (1978a).

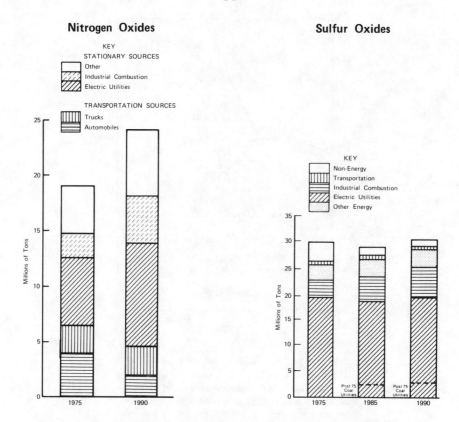

FIGURE 4.6   Net emissions of nitrogen oxides and sulfur oxides for the United States by source for 1975 and projections for 1990. Projections are based on 1979 emissions regulations and national energy use projections. SOURCE: Mitre Corporation (1979).

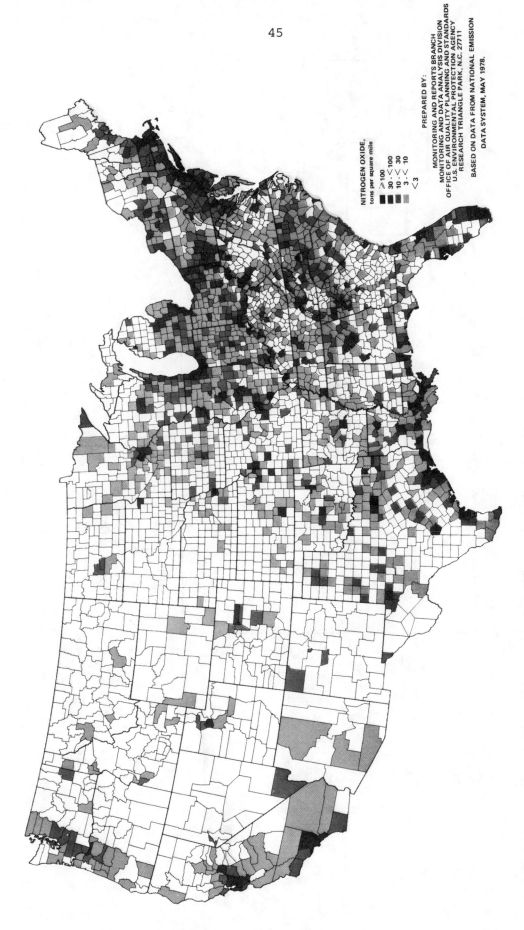

FIGURE 4.7 Nitrogen oxide emission density in the United States by county. SOURCE: U.S. EPA (1978a).

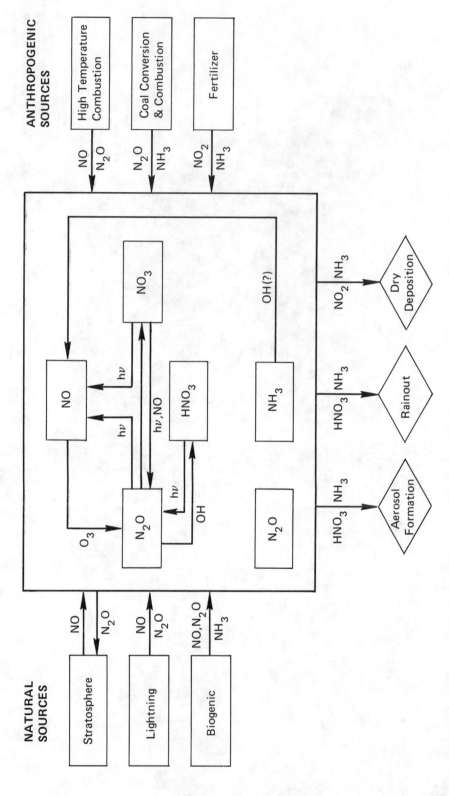

FIGURE 4.8 Nitrogen chemistry in the troposphere. SOURCE: NASA (1981).

But many of the details of the sources, transport, sinks, and reaction kinetics of most tropospheric trace nitrogen species are incompletely understood.

The major anthropogenic source of NO_x (NO and NO_2) in the United States is fossil fuel combustion, and a decline in NO_x emissions projected for automobiles will more than be offset by increased emissions from stationary sources (Figure 4.6). By 1985 stationary sources are expected to account for 70 percent of anthropogenic NO_x emissions (US EPA 1980). Trends in total NO_x emissions have been strongly upward, with almost a three-fold increase over the past 25 years (Table 4.2). Significant increases in NO_x emissions are forecast for the remainder of the century, primarily due to emissions associated with increased coal use in the electric utility and industrial sectors of the U.S. economy.

Nitrogen oxides produced by fossil fuel combustion can create local pollutant levels that are 10 to 100 times greater than natural. However, the regional and global significance of anthropogenic NO_x emissions remains a major problem area for research. Recent estimates of global NO_x production by lightning range from 1.8×10^{12} g to 18×10^{12} g of nitrogen per year (Chameides et al. 1977, Levine et al. 1981). If the lower value is correct, NO_x emissions from anthropogenic sources, estimated to be at least 20×10^{12} g of nitrogen per year, may be a major source of NO_x to the global troposphere (Levine et al. 1981). Biomass burning is also postulated to be an important source of NO_x, with a calculated potential source strength of 20 to 100×10^{12} g of nitrogen per year (Crutzen et al. 1979). Clearly, reducing the uncertainty in estimates of natural sources of NO_x is fundamental to an assessment of the significance of anthropogenic NO_x emissions on regional and global air quality.

NO_x emissions are hypothesized to affect the biosphere through human health effects (NRC 1977c) and through conversion to nitric acid, which may contribute substantially to the acid rain problem (NRC 1978d). These, and other problems related to the nitrogen cycle, can only be resolved by increased research on the biogeochemical cycling of nitrogen oxides.

Trace Metals

Data on emissions of trace metals are very limited, and there is an urgent need for quantitative data on specific rates of emission for both industrial processes and natural volatilization from soils and vegetation. Goldberg (1976) and Harriss and Hohenemser (1978) have discussed emission factors for mercury, for example, but critical difficulties are encountered in obtaining data on the mercury content of certain fossil fuels and raw materials used for metals production and, more importantly, in rates of volatilization from natural surfaces. A summary of various attempts at quantification of anthropogenic mercury emissions is illustrated in Table 4.3. The range in these estimates demonstrates that uncertainties over anthropogenic mercury emissions are at least one order of magnitude.

TABLE 4.2   Total U.S. NO$_x$ Emissions for the Years 1950 and 1975 ($\times 10^{12}$ g/year)

	1950	1975
Utility combustion	1.1	6.1
Other combustion	3.7	5.5
Non-ferrous smelters	neg.	neg.
Other industrial processes	0.3	0.7
Transportation	3.0	9.9
Total	8.1	22.2

NOTE: Values are prorated on the basis of five categories of emission sources. For 1975 the U.S. point sources are at their 1977–1978 emission rate, whereas area sources are at their 1973–1977 emission rate.

SOURCE: U.S.–Canada Research Consultation Group on LRTAP (1979).

TABLE 4.3   Range in Estimates of Anthropogenic Mercury
Emissions to the Global Atmosphere ($\times 10^8$ g/year)

Emission Source	Mercury Flux Estimates[a]
Coal combustion	0.017–63.8
Oil and gas combustion	0.6–22.9
Metals refining	7.3–20.0
Chlor-alkali production	2.6–30.0
Cement manufacturing	1.0–1.3
Total emissions	11.5–137.4

[a]The range in global mercury emission fluxes were obtained
from a review of Bertine and Goldberg (1971), Garrels et al.
(1975), Joensuu (1971), Harriss and Hohenemser (1978),
NRC (1978c), and Lantzy and Mackenzie (1979).

Uncertainties regarding natural sources of mercury are much worse, thus we cannot begin to make reasonable comparisons of anthropogenic and natural mercury sources at present. And the data base on atmospheric emissions for other trace metals and even for most compounds is poorer than that for mercury, indicating the serious problems with uncertainty in making emissions inventories with such limited data.

Estimates of the fluxes of metals to the atmosphere from fossil fuel burning have been made by Bertine and Goldberg (1971), using figures for the consumption of fossil fuels from 1967: 1.75×10^{15} g of coal; 1.04×10^{15} g of lignite; 1.63×10^{15} g of oil, and 0.66×10^{15} g of natural gas. The literature was surveyed for reasonable values of the elemental contents of these fuels, and the authors assumed that the fly ash released to the atmosphere from the burning of coals and oils is about 10 percent of the total ash and that 50 percent of the coal is used in the manufacture of coke. The results are given in Table 4.4. For such elements as barium and mercury, mobilization from fossil fuels to the atmosphere appears to be within an order of magnitude of the river fluxes of these metals to the oceans--which is to say that society has become an important geological agent.

For certain elements these estimates may be low, because volatilization may be selective. Emission spectrographers have noted that there is a preferential volatility of some elements such as arsenic, mercury, cadmium, tin, antimony, lead, zinc, thallium, silver, and bismuth under the high temperature conditions they employ in the direct current electric arc. The fluxes of these metals given in Table 4.4 may be underestimated because they do not take this preferential volatility into account.

There are few data on emissions to the atmosphere from human activities other than fossil fuel use. Nriagu (1979) has estimated the emissions of 5 metals--cadmium, copper, lead, nickel, and zinc--from a variety of anthropogenic activities and sources. These estimates are given in Table 4.5 along with the natural worldwide flux for these 5 trace metals. The data presented for trace metal fluxes from coal and oil combustion differ somewhat from those of Bertine and Goldberg given in Table 4.4, but in any case, anthropogenic emissions to the atmosphere exceed natural emissions for these 5 metals.

Similar fluxes of some trace metals to the atmosphere appear to result from cement production, which equaled 5.7×10^{14} grams/year in 1972 (Goldberg 1976). About 95 percent of this output is portland cement, whose chemical formulation can be considered as one-third shale and two-thirds limestone. Cement is produced by roasting such a mixture at temperatures between 1,450° and 1,600°C, for 2 to 4 hours. Metals whose oxides have boiling points below 2,000°C can be expected to be volatilized.

Hypothesized mobilization of some trace metals to the atmosphere as a result of cement production is given in Table 4.6. Vaporization by this process appears greater than that by fossil fuel burning for such elements as arsenic, boron, lead, selenium, and zinc. This may be true for other metals for which data on boiling points are not at

TABLE 4.4 Amounts of Elements Mobilized into the Atmosphere as a Result of Weathering Processes and the Combustion of Fossil Fuels

| Element | Fossil Fuel Concentration (ppm) | | Fossil Fuel Mobilization ($\times 10^9$ g/year) | | | Weathering Mobilization ($\times 10^9$ g/year) | |
	Coal	Oil	Coal	Oil	Total	River Flow	Sediments
Li	65		9			110	12
Be	3	0.0004	0.41	0.00006	0.41		5.6
B	75	0.002	10.5	0.0003	10.5	360	
Na	2,000	2	280	0.33	280	230,000	57,000
Mg	2,000	0.1	280	0.02	280	148,000	42,000
Al	10,000	0.5	1,400	0.08	1,400	14,000	140,000
P	500		70			720	
S	20,000	3,400	2,800	550	3,400	140,000	
Cl	1,000		140			280,000	
K	1,000		140			83,000	48,000
Ca	10,000	5	1,400	0.82	1,400	540,000	70,000
Sc	5	0.001	0.7	0.0002	0.7	0.14	10
Ti	500	0.1	70	0.02	70	108	9,000
V	25	50	3.5	8.2	12	32	280
Cr	10	0.3	1.4	0.05	1.5	36	200
Mn	50	0.1	7	0.02	7	250	2,000
Fe	10,000	2.5	1,400	0.41	1,400	24,000	100,000
Co	5	0.2	0.7	0.03	0.7	7.2	8
Ni	15	10	2.1	1.6	3.7	11	160
Cu	15	0.14	2.1	0.023	2.1	250	80
Zn	50	0.25	7	0.04	7	720	80
Ga	7	0.01	1	0.002	1	3	30
Ge	5	0.001	0.7	0.0002	0.7		12
As	5	0.01	0.7	0.002	0.7	72	
Se	3	0.17	0.42	0.03	0.45	7.2	
Rb	100		14			36	600
Sr	500	0.1	70	0.02	70	1,800	600
Y	10	0.001	1.4	0.0002	1.4	25	60
Mo	5	10	0.7	1.6	2.3	36	28
Ag	0.5	0.0001	0.07	0.00002	0.07	11	0.03
Cd		0.01		0.002			0.5
Sn	2	0.01	0.28	0.002	0.28		11
Ba	500	0.1	70	0.02	70	360	500
La	10	0.005	1.4	0.0008	1.4	7.2	40
Ce	11.5	0.01	1.6	0.002	1.6	2.2	90
Pr	2.2		0.31			1.1	11
Nd	4.7		0.65			7.2	50
Sm	1.6		0.22			1.1	13
Eu	0.7		0.1			0.25	2.1
Gd	1.6		0.22			1.4	13
Tb	0.3		0.042			0.29	
Ho	0.3		0.042			0.36	2.3
Er	0.6	0.001	0.085	0.0002	0.085	1.8	5.0
Tm	0.1		0.014			0.32	0.4
Yb	0.5		0.07			1.8	5.3
Lu	0.07		0.01			0.29	1.5
Re	0.05		0.007				0.001
Hg	0.012	10	0.0017	1.6	1.6	2.5	1.0
Pb	25	0.3	3.5	0.05	3.6	110	21
Bi	5.5		0.75				0.6
U	1.0	0.001	0.14	0.001	0.14	11	8

SOURCE: Bertine and Goldberg (1971). Bertine and Goldberg used values from a variety of sources discussed in their paper. Reprinted with permission from *Science* 172:233–235. Copyright ©1971 by the American Association for the Advancement of Science.

TABLE 4.5 Worldwide Anthropogenic and Natural Emissions of Trace Metals During 1975

	Global Production or Consumption (×10^{12} g/year)	Trace Metal Emissions (×10^9 g/year)				
		Cd	Cu	Pb	Ni	Zn
Anthropogenic emissions						
Mining, nonferrous metals	16	0.002	0.8	8.2		1.6
Primary nonferrous metal production						
Cd	0.0017	0.11				
Cu	7.9	1.6	19.7	27	1.5	6.6
Pb	4.0	0.20	0.29	31	0.34	0.44
Ni	0.8			2.5	7.2	0.68
Zn	5.6	2.8	0.78	16	0.36	99
Secondary nonferrous metal production	4.0	0.60	0.33	0.77	0.2	9.5
Iron and steel production	1300	0.07	5.9	50	1.2	35
Industrial applications	–	0.05	4.9	7.4	1.9	26
Coal combustion	3100	0.06	4.7	14 '	0.66	15
Oil (including gasoline) combustion	2800	0.003	0.74	273	27	0.07
Wood combustion	640	0.2	12	4.5	3.0	75
Waste incineration	1500	1.4	5.3	8.9	3.4	37
Manufacture, phosphate fertilizers	118	0.21	0.6	0.05	0.6	1.8
Miscellaneous	–	–	–	5.9	–	6.7
Total anthropogenic		7.3	56	449	47	314
Natural emissions[a]		0.83	18.5	24.5	26	43.5

[a]Includes estimates for wind blown dust, forest fires, vulcanism, vegetation, and sea spray.

SOURCE: After Nriagu (1979).

TABLE 4.6 Emissions of Volatile Oxides from the Production of Cement

Element	Boiling Point of Oxide	Parts per Million in Shales	Parts per Million in Limestones	Grams in 5.7 × 10^{14} Grams of Cement	Emission in Grams per Year
Sb	1550	1.5	0.2	4.2×10^8	2.1×10^8
As	315	13.	1.	3.2×10^9	3.2×10^9
B	~1860	100.	20.	3.2×10^{10}	3.3×10^{10}
Cd	decompresses 900-1000	0.3	0.035	8.1×10^7	8.1×10^7
Ca	690 (metal)	6	6	5.2×10^9	5.2×10^9
Pb	1744 (metal)	20	9	9.9×10^{10}	3.0×10^{10}
Li	1200^{600} mm Hg (metal)	66	5	1.4×10^{10}	1.4×10^{10}
Hg	356 (metal)	0.4	0.04	1.0×10^8	1.0×10^8
Rb	decompresses 400	140	3	2.0×10^{10}	2.9×10^{10}
Se	sublimes 350	0.6	0.88	7.1×10^8	7.1×10^8
Tl	1457 (metal)	1.4	0.0	2.7×10^8	2.7×10^8
Zn	907 (metal)	95.	20.0	3.2×10^{10}	3.2×10^{10}

SOURCE: Goldberg (1976).

present available. In some cases where the oxides decomposed upon heating below 2,000°C, the volatility of the metal was used. The calculated emissions for Pb, Cd, and Zn from cement production given in Table 4.6 compare satisfactorily with the emissions for these metals given in Table 4.5 for industrial applications. Such a comparison assumes that the fluxes listed for industrial applications result primarily from cement production, and also takes into account the fact that there may be a systematic bias for high fluxes in the cement production model used to calculate the values given in Table 4.6.

Organic Compounds

Anthropogenic emissions of organic compounds have received substantial attention because of their effect on air quality in urban areas. The emphasis has been on gaseous reactive hydrocarbons, which lead to photochemical smog including high ozone concentrations, and on particulate polycyclic compounds that have mutagenic and carcinogenic properties. This section summarizes current data on anthropogenic organic emissions and briefly discusses the possible effect of new energy technologies on the nature of these emissions.

Global anthropogenic emissions of non-methane hydrocarbons are estimated to be approximately 80×10^{12} g per year (Duce 1978). Duce derived this emission rate from the earlier estimates of Robinson and Robbins (1968) and more recent data from the U.S. Environmental Protection Agency (1976) for the United States. These emissions originate primarily from fossil fuel burning. Although global anthropogenic emissions of gaseous organics are approximately one order of magnitude lower than natural emissions, emissions are mostly confined to the industrialized regions of the Northern Hemisphere and have significant effect on air quality in these regions.

Anthropogenic emissions of particulate organic carbon were estimated by Duce (1978) to be approximately 30×10^{12} g per year in 1973-74 period. Duce broke down emissions data by type of industrial source and separated particulates into two size fractions, greater than and less than 1 micron diameter (Table 4.7). Four major sources--coal, petroleum, noncommercial fuel, and agricultural burning--account for approximately 80 percent of the total anthropogenic emissions of organic carbon particles. The molecular composition of these anthropogenic emissions has been reviewed (e.g. Grosjean 1977, Graedel 1978). Because of their adverse health effects, the polycyclic aromatic hydrocarbons (PAH) have received considerable attention (NRC 1972), and specific emission factors for anthropogenic sources of benzo(a)pyrene and other PAH are now available (U.S. EPA 1978b).

Both increased emissions of existing organic pollutants and emissions of new pollutants may result from emerging fossil fuel technologies (coal gasification and liquefaction, shale oil and tar sand exploitation) and from increasing use of diesel and gasoline-alcohol ("gasohol") fuels. Moderate to large increases in

TABLE 4.7 Estimates of Global Primary Particulate Organic Carbon (POC) Emissions from Anthropogenic Sources, 1973-1974

Source	World Production or Consumption ($\times10^{12}$ g/year)	Emission Factor (tons/ton)	Total Particulate Emissions ($\times10^{12}$ g/year)	Estimated Percentage Carbon	POC Emissions ($\times10^{12}$ g/year)		
					d > 1μm	d < 1μm	Total
Coal	3069						
Power production	1290	0.011	14.2	20	1.4	1.4	2.8
Industry	770	0.026	20.0	20	2.0	2.0	4.0
Domestic and commericial fuel	400	0.01	4.0	20	0.4	0.4	0.8
Cleaning refuse	860	0.0021	1.8	20	0.2	0.2	0.4
Coke	910	0.0026	2.4	20	0.2	0.2	0.4
Carbon black	4.2	0.083	0.3	80	0.1	0.1	0.2
Cement	696	0.012	6.5	5	0.3	–	0.3
Pig iron and crude steel	1220	0.009	11.0	5	–	0.6	0.6
Ferroalloys	12	0.06	0.7	5	–	–	–
Copper smelting	8.7	0.17	1.5	5	–	0.1	0.1
Al, Pb, and Zn products	21.7	0.06	1.2	5	0.2	0.1	0.1
Lime production	119	0.032	3.8	5	–	–	0.2
Nitric acid production	30	0.008	0.2	5	–	–	–
Phosphate fertilizer	23	0.05	1.2	25	0.3	–	0.3
Chemical wood pulp	91	0.01	0.9	40	0.2	0.2	0.4
Incineration	630	0.0085	5.4	20	0.9	0.2	1.1
Noncommercial fuel	1940	0.005	9.7	65	1.9	4.4	6.3
Cotton ginning	14	0.011	0.2	40	0.1	–	0.1
Wheat handling	360	0.025	9.0	40	3.6	–	3.6
Petroleum refining	2650	0.00014	0.4	80	–	0.3	0.3
Petroleum combustion							
Gasoline	613	0.0019	1.2	80	0.2	0.8	1.0
Kerosene	78	0.0010	0.1	80	–	0.1	0.1
Fuel oil	592	0.0010	0.6	40	0.1	0.2	0.3
Residual oil	959	0.0014	1.3	40	0.1	0.4	0.5
Aircraft jet fuel	106	0.00015	–	–	–	–	–
Natural gas	1.3×10^{12} m³	0.28 t/10^6 m³	0.4	80	0.1	0.3	0.4
Agricultural burning	1000	0.0085	8.5	65	1.6	3.9	5.5
Total					13.9	15.9	29.8

SOURCE: Duce (1978). Reprinted with permission from *Pure and Applied Geophysics* 116:244.

particulate PAH emissions are expected to result from increased use of coal, fuels from coal gasification and liquefaction, and shale-derived fuels; use of diesel-powered vehicles (NRC 1981b); and burning of wood for domestic heating (e.g., Butcher and Sorenson 1979).

Organic pollutants other than PAH that may require attention include phenols from coal process waters (Guerin 1977), aldehydes in diesel-powered vehicle exhaust (NRC 1981b), and unburned alcohol as well as aldehydes in exhaust from gasohol-powered vehicles (Allsup and Eccleston 1980). In addition, new heteroatomic organic pollutants including sulfur-containing and nitrogen-containing compounds may be emitted to the atmosphere as a result of fuel conversion processes. These organic gases include mercaptans, sulfides, thiophenes, furans, pyrroles, and pyridines (Sickles et al. 1977). The environmental persistence and fate of these compounds have received little or no attention to date.

SUMMARY

Fossil fuel burning, primarily of coal and oil, contribute most of the anthropogenic constituents of the atmosphere, and the bulk of the burning is done in the Northern Hemisphere, primarily in North America, Europe, and Japan. The emerging fuel technologies (coal gasification, shale oil, tar sands, etc.) as well as the increased use of diesel and alcohol fuels may alter the magnitude and character of pollutant fluxes to the atmosphere in the future.

Gaseous pollutants include sulfur dioxide, nitrogen oxides, organic compounds, and trace metals. After entering the atmosphere they may be oxidized (sulfur dioxide going to sulfate, the nitrogen oxides going to nitrate), associate with aerosols, or be degraded. Almost without exception they return quantitatively to the earth's surface.

Trace metals from fossil fuel combustion or cement production may initially enter the atmosphere as gases but, in general, quickly become associated with the aerosols or form aerosols. Both fossil fuel combustion and cement production are introducing some metals to the atmosphere at rates comparable to river fluxes to the oceans, showing that such activities have made society an important geological agent.

The particulate organic carbon releases that come from coal, petroleum, noncommercial fuel, and agricultural burning account for 80 percent of the fluxes. Of particular concern are the polycyclic aromatic hydrocarbons, which pose a direct health hazard.

5. ATMOSPHERIC TRANSPORT, TRANSFORMATION, AND DEPOSITION PROCESSES

The atmosphere serves as the delivery system from emission sources to the biosphere. Once substances are in the atmosphere, what happens to them depends on their physical and chemical characteristics rather than on whether they were of natural or anthropogenic origin. This chapter describes briefly the kinds of processes all substances undergo once they have entered the atmosphere--the processes of transport, diffusion, chemical transformation, and deposition.

The processes substances undergo in the atmosphere, their pathways into the biosphere, and their atmospheric residence times depend upon such characteristics as physical state, particle size, and chemical reactivity. Figure 5.1 gives examples of atmospheric constituents with various residence times and the corresponding distances they may typically be transported. In general, properties conducive to short atmospheric lifetimes are large particle size and high reactivity and solubility. For example, large aggregated particles containing lead from automobile exhausts are deposited in close proximity to the roadways where they are emitted, and large particles emitted from smelters (e.g., Cu, Fe, Ni, Zn) typically remain in the atmosphere for at most a few hours and to a large extent fall out within a few tens of miles from their source (Jeffries and Snyder 1980). Substances that are highly reactive--e.g., NO, HCl, SO_2, HNO_3--and/or have a high solubility in water--e.g., NH_3, SO_2--or hygroscopicity--e.g., aerosols of NH_4NO_3 and $(NH_4)_2SO_4$--are fairly rapidly removed; that is, their residence times are on the order of hours to a week. Substances that remain as small particles (i.e., in the submicrometer range) or that are sparingly soluble (e.g., Pb, V, Hg) may have residence times of several weeks and more. These reactive gases and small particles frequently undergo several transformation and removal processes in the atmosphere before finally being deposited at the earth's surface and thus have a rather complex pathway to the biosphere.

The efficiency with which the less reactive substances such as some vapor-phase organic pollutants are removed from the atmosphere depends upon both their volatility and the rate at which they are adsorbed on solid particles in the atmosphere. Radon and

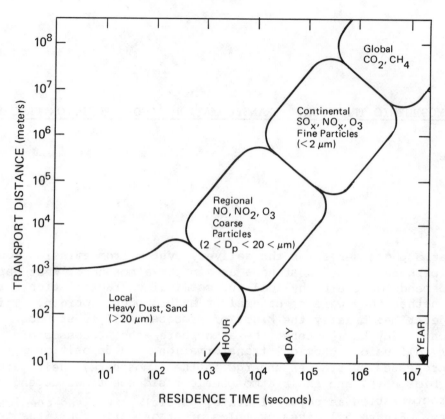

FIGURE 5.1 Dispersion of pollutants introduced into the atmosphere as determined by residence time. Man-made sulfur compounds, including fine particles, are distributed on a continental scale. SOURCE: R. B. Husar and D. E. Patterson, Center for Air Pollution Impact and Trend Analysis, Washington University, St. Louis, Missouri, personal communication, 1980.

organochlorines, which are both gaseous and nonreactive, remain in the atmosphere for several years and become distributed throughout the global troposphere and stratosphere.

Because the atmospheric residence times and transport distances of many pollutants are large, substances introduced to the atmosphere from one country or region can significantly affect another. There are many such examples. Sulfur dioxide emitted from fossil fuel burning plants in central Europe and North America is known to be deposited hundreds of kilometers from its source. Radioactive debris from the explosions of a Chinese nuclear device detonated in May 1965 was detected at sampling sites in Tokyo and Fayetteville, Arizona: the average velocity of the wind transport was about 16 m/sec in the tropospheric jet streams and two circumnavigations of the world were evident from fallout in both June and July 1965. Part of the DDT sprayed upon agricultural crops in Africa precipitates detectably in the Caribbean. Organic pollutants such as PCBs, hexachlorobenzene, and pesticides, have been detected in the atmosphere over remote parts of the oceans (Atlas and Giam 1981).

In examining atmospheric processes, it is convenient to group pollutants in terms of their residence time or zone of influence. In focusing on atmosphere-biosphere interactions, however, it may be more fruitful to use characteristics such as solubility and reactivity, which are consistent with atmospheric groupings; or to use ecological toxicity, synergistic relationships, and persistence, which are not.

TRANSPORT AND DIFFUSION

Pollutants are mixed or dispersed through the lower atmosphere by turbulent diffusion, vertical wind shear, and precipitation. Turbulence is generated both mechanically--e.g., by wind interaction with the surface, changes in surface roughness, and wind shear--and thermally--e.g., by from solar radiation heating the underlying surface and generating convective motions. The larger the scale and intensity of turbulence, the more efficient the mixing. Wind shear--change in wind direction with height--causes a horizontal spreading of pollutants. Rain falling through an SO_2 plume can cause a vertical redistribution of pollutants.

The vertical structure of the lower troposphere is important to pollutant transport. Wind speed increases with height as the effect of surface roughness diminishes. Thus, the higher a pollutant's effective injection height (stack height plus plume rise), the greater the transport wind speed. In general, atmospheric temperature decreases with height above the surface; the actual variation of temperature above the surface at a given time and place defines the stability of the atmosphere and thus the amount of vertical mixing. Pollutants emitted into an unstable layer are mixed throughout the layer; pollutants emitted into a stable atmospheric layer are mixed very little in the vertical dimension.

The structure of the near-surface layer of the atmosphere, the planetary boundary layer, varies on a diurnal schedule that affects pollutant transport and diffusion. At night, the loss of heat as long-wave radiation cools the land surface and in turn the near-surface air, causing a ground-based stable layer to form--a condition known as inversion. Pollutants emitted into this layer undergo little mixing or dilution, while those emitted above it may be slowly mixed through a higher layer of the atmosphere above the cool, ground-based layer without reaching the surface. In the morning, as solar radiation heats the surface and causes convective mixing, the stable layer is eroded from below and pollutants mix progressively higher in the atmosphere, frequently up to 1 or 2 km, depending on the time of year and meteorological conditions. The following night, the cycle is repeated: pollutants well mixed from the previous day remain above the newly formed surface layer, and new pollutants are injected into the lower, stable layer. Although this picture is rather simplified and pertains primarily to fair-weather conditions in nonpolar, continental areas, it does indicate the complexity of the atmospheric processes controlling pollutant behavior and the difficulty in modeling these processes.

Pollutants can be transported under a variety of meteorological conditions. Frequently, however, for lack of time and space resolution in meteorological and pollutant measurements, transport of pollutants is envisioned as occurring at the average speed of the wind in a layer through which the pollutants are assumed to be uniformly mixed.

Plumes emitted into a stable atmosphere undergo little vertical or horizontal diffusion and can travel intact for several hundred kilometers before being dispersed or incorporated into cloud. Figure 5.2 shows such a narrow, coherent plume observed by satellite over Ontario, Canada. On the other hand, when emissions from diverse sources over a broad area accumulate in stagnating air associated with anticyclonic conditions of eastern North America or western Europe, the pollutants become well mixed by daytime convection and are slowly transported in the southerly flows to the west of the high pressure centers, to affect areas several hundreds of km across for several days at a time. A related transport situation is responsible for many of the episodes of excessive sulphate and hydrogen ion deposition in Scandinavia. In stagnant anticyclonic conditions over Europe, the air becomes heavily polluted and is then drawn into the frontal area of a depression running along the northern edge of the anticyclone. Only modest rainfall at the front is needed to produce relatively large deposition of atmospheric contaminants.

TRANSFORMATIONS

In terms of chemistry, prediction of end products and concentrations of inorganic pollutants would appear to be very straightforward. Owing to the abundance of oxygen in the atmosphere and its high reactivity, most elements will tend to form oxides of various sorts. The calculated equilibrium partial pressures for reduced gases such as CH_4, NH_3, and H_2S are ridiculously small.

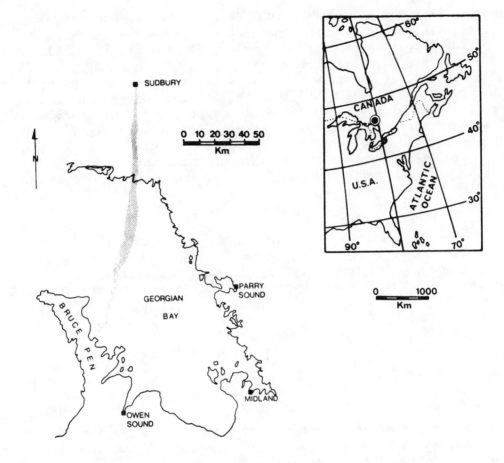

FIGURE 5.2 Tracing from ERTS photograph showing the outline of a plume from the 381-m nickel smelter stack at Sudbury, Ontario, crossing Georgian Bay, 1040 EST, September 1972. The inset shows the map location in central Ontario. SOURCE: After Munn (1976).

For example, at chemical equilibrium, there would be less than one H_2S molecule in the entire atmosphere! Measurements have shown that the true concentration is over 1033 molecules and that the oxidation of H_2S to sulfate aerosols occurs within about 12 hours of its emission (Jaechke et al. 1978). It is clear that the atmosphere is not at chemical equilibrium and that the enormous disequilibrium we observe does not simply represent a slow approach to equilibrium. Chemical disequilibrium for H_2S is maintained in large measure by sulfate-reducing microorganisms. Biological processes are closely coupled with the atmosphere: the state of the atmosphere is as profoundly altered by life on the planet as biological processes are affected by the atmosphere.

Oxidation Reactions in the Atmosphere

Let us now examine the preceding generalities as they operate in the atmosphere. Two initiating photochemical reactions are especially important:

[1] $$NO_2 + h\nu \longrightarrow NO + O(^3P)$$

[2] $$O_3 + h\nu \longrightarrow O_2 + O(^1D)$$

where $h\nu$ signifies a quantum of energy (photon), and 3P and 1D signify the electron excitation state of the oxygen atom. Each of these reactions is followed by ozone synthesis:

[3] $$O + O_2 + M \longrightarrow O_3 + M$$

where M represents a third body in the collision of reactant atoms and molecules. Subsequent free-radical chemistry produces very reactive hydroxyl free radicals by chain processes, which result in many free radicals being produced for every primary photon:

[4] $$O(^1D) + H_2O \longrightarrow OH + OH$$

[5] $$OH + O_3 \longrightarrow OH_2 + O_2$$

[6] $$OH_2 + NO \longrightarrow NO_2 + OH$$

with the NO_2 recycling to equation (1), above, and

[7] $$OH + H_2S \longrightarrow HS + H_2O$$

[8] OH + SO_2 ⟶ sulfate aerosols

[9] OH + CH_4 ⟶ H_2O + CH_3

[10] CH_3 $\xrightarrow{O_2,\ NO,\ OH}$ H_2CO $\xrightarrow{h\nu,\ OH}$ CO

[11] OH + CO ⟶ CO_2 + H

[12] OH + NO_2 ⟶ HNO_3 (vapor)

Free radicals, such as the hydroxyl (OH) and peroxyl (OH_2) fragments in the reactions above, are both thermodynamically reactive, in that their unpaired electrons can be thought of as half of yet unborn exothermic bonds, and kinetically reactive, in that the half bonds hold the fragments together weakly and are thus easily rearranged by thermal collisions. In the atmosphere, the hydroxyl free radical acts ubiquitously as a kinetic messenger carrying the "memory" of the initiating solar photons and facilitating subsequent reactions, such as the sulfur oxidations in equations 7 and 8 and the methane reactions schematically abbreviated in equations 9 through 11. The chemistries of such processes are relatively simple. The primary photosteps (equations 1 and 2) drive what would otherwise be endothermic reactions to dissociate more organized chemical species (O_3 and NO_2) into less organized fragments (O, O_2, and NO); subsequent exothermic chemistry (equations 3 through 12) produces more organized species, such as the sulfate aerosols. The chemical potential driving the chemistry toward equilibrium is affected by both the energy transfer, which is called the change in enthalpy, H, and the reorganization, which is called the change in entropy, S.

The atmosphere, then, is not at equilibrium; the observed levels of atmospheric constituents are controlled by reaction rates, especially those modulated through the OH free radical.

Reduction Reactions in the Biosphere

The chemical potentials for oxidation of reduced atmospheric species such as hydrogen sulfide and methane are very negative--that is, oxidation of these species are strongly favored. For reduced species to be formed at all, they must be formed in anoxic environments where the oxygen partial pressure is sufficiently low to permit a finite partial pressure of the reduced gas as an activity product in the denominator of equilibrium ratios such as:

$$K_P = \frac{P_{H_2O}}{P_{O_2}} \cdot \left[\frac{P_{CO_2}}{P_{CH_4}} \right]^{\frac{1}{2}}$$

where K_p is a dimensionless ratio and P values are the partial pressure for those gases.

Methane, for example, which is observed at 1 ppmv in the atmosphere, could only exist in equilibrium with the water of the ocean and the atmospheric concentration of CO_2 at 330 ppmv if the partial pressure of oxygen were 4×10^{-73} atmospheres. Because methane is found in the atmosphere in the presence of O_2 at pressures near 0.2 atmospheres, methane synthesis clearly must occur in sites that are shielded from oxygen.

The atmosphere, then, is not considered an isolated reservoir. The oxidation of methane in the atmosphere (equations 9 through 11) displays a residence time of about 4 years. The total atmospheric inventory of methane is about 10^{38} molecules. Thus a flux of methane into the atmosphere of about 1×10^{15} g per year is required to maintain the observed concentration of methane. This flux must originate in sites closely coupled with the atmosphere but capable of maintaining a very low oxygen partial pressure. It is the biogenesis of methane by microorganisms, plants, and animals that maintains this flux.

Organic Compounds

Chemical reactions of organic species in the "clean" troposphere have not been extensively studied, and most of our knowledge concerning these reactions derives from studies of urban polluted air (e.g. Grosjean 1977, Graedel 1978, Atkinson et al. 1979). The residence times and fates of organic compounds in the troposphere are controlled to a large extent by their chemical reactivity. Gaseous organics undergo complex photooxidation processes leading to the formation of oxygenated products, a fraction of which accumulate in aerosol droplets, where they may react further via heterogeneous oxidation processes. Organic compounds emitted directly into the atmosphere in the particulate phase may also undergo transformations by reaction with oxidizing pollutant gases.

Atmospheric residence times for gaseous organic species may be defined as:

$$\tau = \frac{1}{\Sigma k_i \chi_i}$$

where χ_i are the concentrations of reactive species such as the hydroxyl radical, ozone, etc., and k_i are the corresponding second-order reaction-rate constants. On the basis of known tropospheric concentrations for OH, ozone, etc., and of the known rate constants for their reactions, it appears that atmospheric lifetimes for non-methane hydrocarbons and other gaseous organics in the atmosphere are controlled to a large extent by their reactions with the hydroxyl radical:

$$\frac{- d\ (organic)}{dt} \sim k_{OH}\ (OH)\ (organic)$$

or:

$$\tau \sim \frac{1}{k_{OH}\ (OH)}$$

Rate constants for reaction of the hydroxyl radical with a large number of gaseous organics, including paraffins, olefins, aromatics, aldehydes, ketones, alcohols, esters, sulfur-containing compounds (e.g., mercaptans, sulfides), and nitrogen-containing compounds (e.g., amines, nitrate esters), have been measured over the past five years (Atkinson et al. 1979). Atmospheric residence times can be readily calculated for these compounds. In the case of olefins, including isoprene and terpenes, reaction with ozone is also important:

$$\frac{- d\ (olefin)}{dt} \sim k_{OH}\ (OH) + k_{O_3}\ (O_3)$$

and should be taken into account when estimating olefin residence times in the atmosphere. For typical atmospheric concentrations of about 10^6 OH radicals per cm^3, most organic compounds, except haloalkanes, have atmospheric lifetimes of less than one month, with the most reactive compounds, including terpenes, being chemically transformed and removed in a matter of hours.

Photooxidation of gaseous organics in the atmosphere, initiated by reaction with the OH radical, involves removal of a hydrogen atom (paraffins), addition on an unsaturated bond (olefins), or both (aromatics). Subsequent reactions, which have been studied in detail (e.g. Carter et al. 1979a,b; Grosjean and Friedlander 1979; Atkinson et al. 1979, 1980), lead to the formation of carbonyls and other mono- and poly-functional oxygenates. The oxygenated products thus formed will either condense as aerosols via gas-to-particle conversion processes (Grosjean 1977) or react further in the gas phase. Duce (1978) suggested, on the basis of global budgets for vapor phase and particulate organic compounds, that gas-to-particle conversion processes may be of major importance in the troposphere, but no experimental studies support or refute this hypothesis. Likewise, Hofmann and Rosen (1980) have suggested that conversion in the stratosphere of gases dimethyl sulfied and carbonyl sulfide to aerosols may be causing rapid increases in aerosols.

Significant among the products of the gas-phase reactions will be CO and H_2. Zimmerman et al. (1978) have estimated that photooxidation of hydrocarbon emissions from vegetation is a major atmospheric source of both carbon monoxide (4 to 13 x 10^{14} g/year) and hydrogen (10 to 35 x 10^{12} g/year). Examples of simplified photochemical pathways are given in Figure 5.3 for a typical paraffin, n-butane (Figure 5.3a); a typical olefin, isoprene (Figure 5.3b); a typical aromatic hydrocarbon, toluene (Figure 5.3c); and for cyclic monoterpenes (Figure 5.3d), derived from the mechanism proposed by Grosjean and Friedlander (1979) for cyclic olefins.

Unlike homogeneous (gas-phase) pathways, heterogeneous reactions involving organic substances have received little attention to date. Heterogeneous pathways that may be important in the atmosphere include oxidation in aerosol droplets by free radicals (OH, OH_2; e.g., Graedel et al. 1975) and dissolved oxidizing species (ozone, hydrogen peroxide, nitric acid), and surface oxidation of organic particles by gaseous pollutants (O_3 and HNO_3). Because no gas-phase mechanism readily accounts for the oxidation of aldehydes to carboxylic acids, which are the observed end products of many hydrocarbon photooxidation pathways, formation of carboxylic acids by one or more of the above heterogeneous processes has been suggested (Grosjean and Friedlander 1979). Other heterogeneous processes that may be of importance in the atmosphere are those involving the reactions of particulate polycyclic aromatic hydrocarbons with pollutant gases including ozone, NO_2, and free radicals to form oxygenated and nitro derivatives (e.g. Pitts et al. 1978). Experimental studies of these heterogeneous processes have been initiated only recently, and the atmospheric importance of these mechanisms is not known.

DEPOSITION

The transfer of trace substances from the atmosphere to surface receptors is accomplished by a variety of physical, chemical, and biological processes. A reasonable qualitative understanding exists of many of the individual mechanisms of transfer, but quantitative knowledge of the phenomena is limited. In many cases, transfer rates, or fluxes between reservoirs, are not well known. In others, the relative importance of the several removal processes acting on a particular substance are not known. These deficiencies arise, in large part, from an inability to measure accurately the mass transfers occurring by several of the deposition processes, particularly dry deposition.

Deposition, or removal, processes are conveniently separated into two categories: those which involve precipitation, called wet deposition processes, and those which do not involve precipitation and may go on all the time, called dry deposition processes. Removal involving fog, mist, and dew lies between these two categories but is closest in character to dry deposition. Either category of process involves both particles and gases.

The effectiveness of individual removal processes depends to a great extent on the physical and chemical characteristics of the particular substance. It should, therefore, be possible to predict

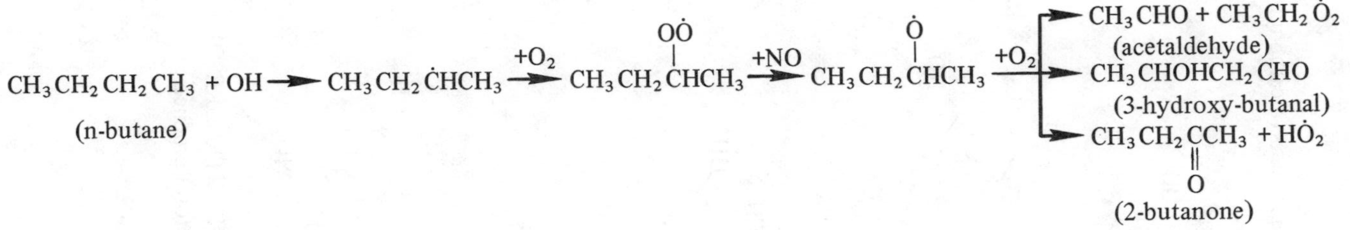

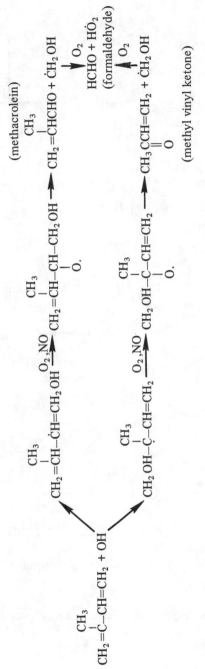

CH₃CH₂CH₂CH₃ + OH → CH₃CH₂ĊHCH₃ $\xrightarrow{+O_2}$ CH₃CH₂CHCH₃ $\xrightarrow{+NO}$ CH₃CH₂CHCH₃ $\xrightarrow{+O_2}$

(n-butane)

CH₃CHO + CH₃CH₂Ȯ₂
(acetaldehyde)
CH₃CHOHCH₂CHO
(3-hydroxy-butanal)
CH₃CH₂CCH₃ + HȮ₂
(2-butanone)

FIGURE 5.3a   Simplified photooxidation mechanisms for n-butane.

CH₂=C—CH=CH₂ + OH

CH₂=CH—ĊH—CH₂OH $\xrightarrow{O_2, NO}$ CH₂=CH—CH—CH₂OH → CH₂=CHCHO + ĊH₂OH
(methacrolein)

CH₂OH—Ċ—CH=CH₂ $\xrightarrow{O_2, NO}$ CH₂OH—C—CH=CH₂ → CH₃CCH=CH₂ + ĊH₂OH
(methyl vinyl ketone)

HCHO + HȮ₂
(formaldehyde)

FIGURE 5.3b   Simplified photooxidation mechanisms for isoprene.

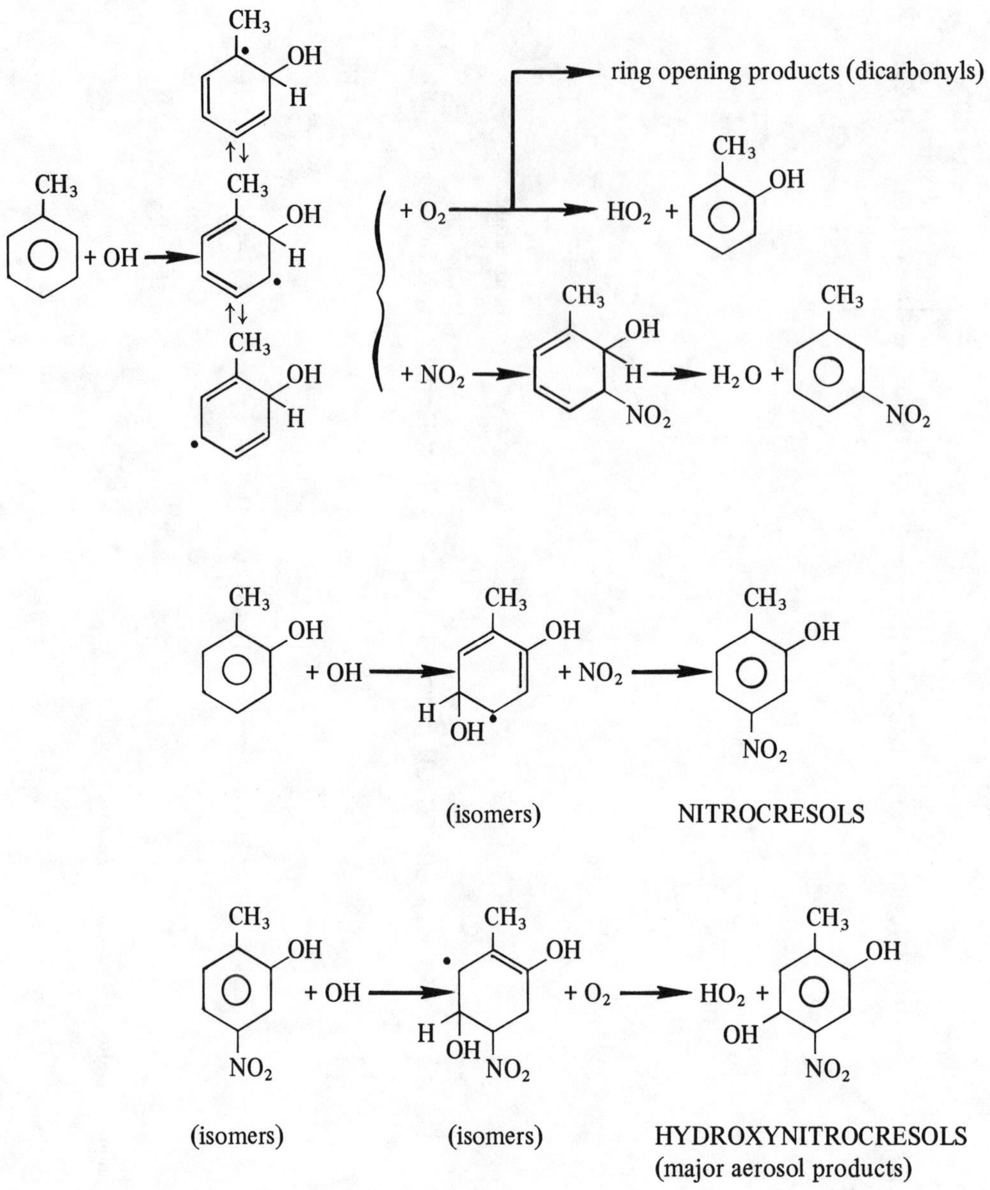

FIGURE 5.3c   Simplified photooxidation mechanisms for toluene.

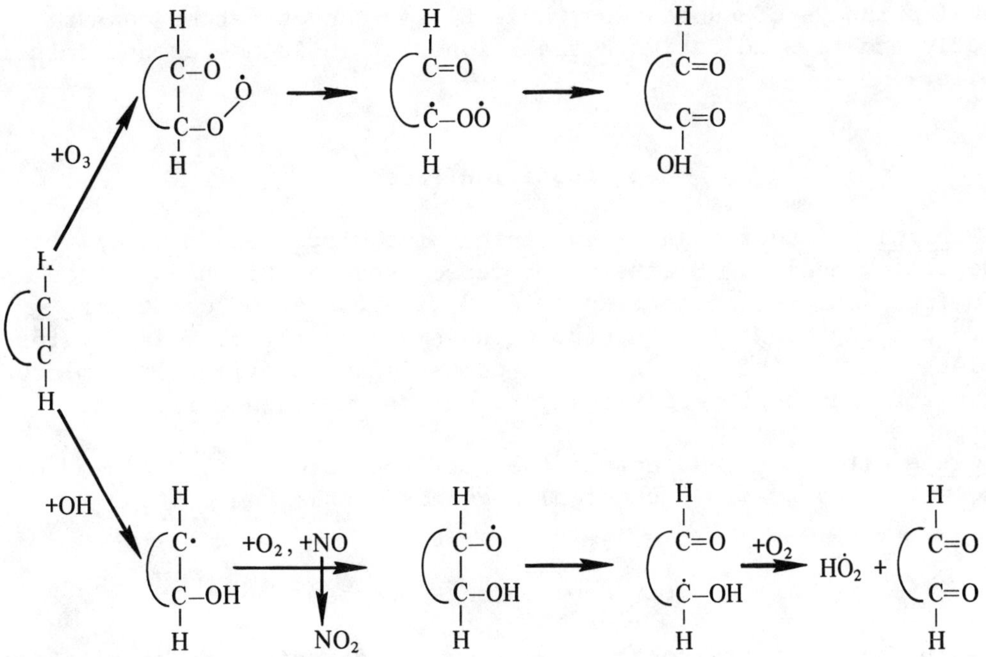

FIGURE 5.3d   Simplified photooxidation mechanisms for cyclic monoterpenes.

removal pathways of substances whose fluxes cannot yet be measured directly and thus to estimate their zones of influence around emission sources.

Wet Deposition Processes

Particles larger than a few tenths of a micron may serve as cloud condensation nuclei. Diffusive processes such as Brownian motion and diffusiophoresis--transport associated with fluxes of condensing water--are important for incorporating smaller particles into water droplets. Below the clouds, large particles--approximately 1 micron or greater--can be intercepted by raindrops and washed from the atmosphere.

Gases diffuse to and across the air/drop interface, followed by dissolution and possibly chemical reaction within the drop.

Dry Deposition Processes

Particles that are relatively large are deposited by gravitational sedimentation. Smaller particles are brought to the surface by turbulent transfer; diffusion through the viscous boundary layer completes the deposition process.

Gases are deposited by turbulent transfer to surfaces followed by diffusive transport through the viscous boundary layer and uptake at the surface by physical adsorption or absorption, biological processes, dissolution, or chemical reaction.

Many elements exist in the atmosphere in more than one compound and phase. For example, sulfur is present as sulfur dioxide, sulfuric acid aerosol, hydrogen sulfide, and other forms. Thus, assessment of the total deposition of sulfur, or even acidic sulfur constituents, necessarily requires the measurement of fluxes by different pathways using different measurement techniques.

The state of the receptor surface may also affect the rate of deposition. Horntvedt et al. (1980) found that when trees were exposed to SO_2 in a wind tunnel, trees with wetted leaves accumulated nearly 100 times more sulfur than trees with dry leaves. Coniferous trees also accumulated much more sulfur than deciduous ones, whether wet or dry (Figure 5.4). Deposition was independent of wind velocity and temperature. Plants exposed for a long period (168 hours) apparently accumulated much less sulfur per unit time than those exposed only half an hour. The authors hypothesized that either depressed photosynthesis and gas exchange or translocation of absorbed SO_2 to the roots might be responsible.

An additional complication in the measurement of deposition is the fact that some substances may be emitted, or deposited and reemitted, from the receptors of interest into the atmosphere. This necessitates either the measurement of upward and downward fluxes separately, or a measurement of net flux. Examples of such substances include mercury, ammonia, and organic compounds such as DDT.

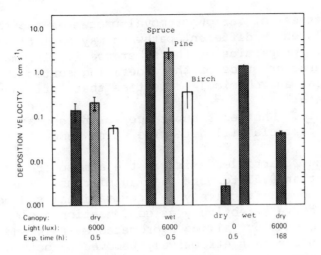

FIGURE 5.4   The effects of wet and dry leaf surfaces and light and dark on the deposition velocity of SO_2 in wind tunnel experiments for three tree species. SOURCE: Horntvedt et al. (1980).

The effectiveness of various removal processes and thus the magnitudes of fluxes by different pathways may be influenced by three sets of factors: properties of the substance itself, meteorological factors, and characteristics of the underlying surface. Some of the important physical and chemical properties that influence removal from the atmosphere are:

Physical state. Whether a substance is present as a gas or as a solid or liquid particle will determine which removal processes are operative.

Particle size. Particles exist in the atmosphere over a range of sizes from less than 0.01 to more than 10 microns. The largest particles are removed effectively by gravitational sedimentation. Large particles serve effectively as condensation nuclei and are scavenged below clouds by falling hydrometeors--rain and snow. Smaller particles are less effectively removed by both wet and dry diffusive processes. Figure 5.5 shows how dry removal of particles depends on size.

Reactivity. The chemical reactivity of a substance contributes to its removal by preventing a buildup of the substance at an interface, which might inhibit further transfer. For example, precipitation is an effective scavenger of sulfur dioxide as long as the composition of the hydrometeors is such as to promote rapid oxidation to sulfate within the hydrometeor. Otherwise, less efficient scavenging occurs.

Surface characteristics of particles. Particles with large surface-to-volume ratios and particles with reactive surfaces may be effective scavengers of other atmospheric constituents--such as vapor-phase organic pollutants (VPOP)--and thus contribute to their removal.

Hygroscopicity. Hygroscopic aerosols are effective condensation nuclei. Their size is significantly affected by relative humidity, which in turn influences the effectiveness with which they are removed by sedimentation or below-cloud scavenging.

Solubility. Soluble gases are readily incorporated into precipitation elements, moist plant surfaces, or surface waters.

Meteorological factors that influence removal include atmospheric stability and intensity of turbulence, which govern the rate of delivery of gases and small particles to the surface, and the frequency, duration, and intensity of precipitation, which determine the relative importance of the wet and dry removal pathways.

For those deposition pathways in which turbulent transfer to the surface is important, the nature of the underlying surface may play a controlling role in the deposition. Surfaces that are rough, hairy, or sticky, such as vegetation, are effective sites for small particle deposition and retention. The surface pH of waters and soils governs the effectiveness of the deposition of some gases, such as carbon dioxide, sulfur dioxide, and perhaps nitrogen oxides, by affecting their reaction rates. The degree of physiological activity--e.g., stomatal opening--can affect the uptake rate of gases and perhaps small particles.

It is useful to think of the dry deposition of gases and small particles in terms of an analogy to resistance (for example, Fowler

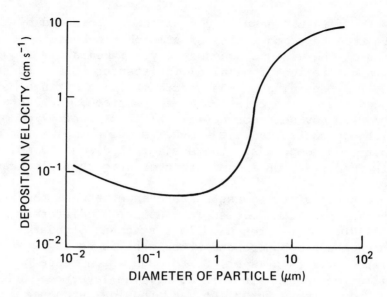

FIGURE 5.5   Velocity of deposition of particles onto short grass.
SOURCE: Chamberlain (1975). Reprinted with permission from
Vegetation and the Atmosphere, volume 1. Copyright © 1975 by
Academic Press, Inc. (London) Ltd.

1980), where the effectiveness of deposition is governed by the total resistance to transfer, consisting of atmospheric resistance, r_a, the resistance of the viscous boundary layer immediately above a surface, r_b, and a series of parallel resistances to the surface, r_c (Figure 5.6). Depending on the nature of the substance being transferred and the characteristics of the surface, the overall deposition process may be atmospheric-resistance controlled or surface-resistance controlled. Table 5.1 shows examples for common sulfur and nitrogen compounds. The subject is discussed more thoroughly in Chapter 6, on accumulation of atmospheric contaminants in the biosphere.

An analogous transfer process for exchange between the atmosphere and water surface has been modeled in a number of different ways (Danckwerts 1970). Most often used is a stagnant boundary layer model, in which the atmosphere and water are viewed as two turbulent bodies separated by a thin layer, Z, through which gases pass by molecular diffusion alone. In this model, the layer becomes thinner as the water becomes more turbulent, and thus gas exchange increases (Kanwisher 1963).

The flux of a gas, F, is calculated as

$$F = E \frac{D}{Z} (C_a - C_l)$$

where

E is an enhancement factor representing chemical
reactivity of the gas in water,
D is the diffusion coefficient for a given gas, and
C_a and C_l are the concentrations of a given gas in
the atmosphere and surface water respectively.

Such models have proved invaluable for measurement of gas exchange in both oceanic and lacustrine environments (Broecker and Peng 1974, Emerson 1975a,b). Although Liss and Slater (1974) have modeled the transfer of SO_2 and other gases, the lack of diffusion coefficients and estimates of chemical reactivity for gases of anthropogenic origin makes the application of such models difficult.

RESIDENCE TIMES FOR SUBSTANCES IN THE ATMOSPHERE

The mode of emission, rate of supply, transformations in the atmosphere, and factors affecting the deposition processes act together to determine the residence time of a substance in the atmosphere, which, in turn, determines how far the substance is likely to be dispersed from its source. Residence time thus is a useful indicator in determining where in the biosphere a given substance will be deposited. Models to estimate residence times have been formulated (see NRC 1978b, Chapter 4, for example). For aerosols, average residence times of the order of ten days are derived. For gases, especially those with low solubilities in water, the residence times

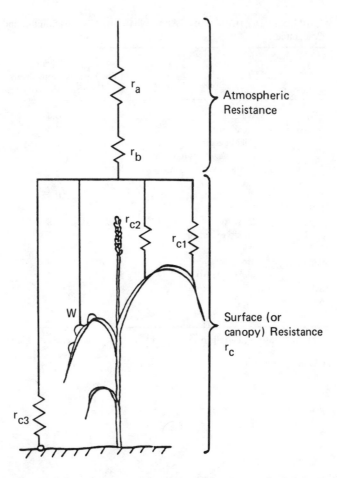

FIGURE 5.6 Resistance to dry deposition of pollutant gases in a cereal crop. Surface resistances are the leaf stomatal component, r_{c1}, the plant cuticular component, r_{c2}, and the soil component, r_{c3}. W refers to the situation with pure water on foliage when normal paths of uptake are short-circuited. The boundary layer resistance, r_b, is in series with the aerodynamic resistance, r_a, in the manner described by Chamberlain (1968). SOURCE: Fowler (1980).

TABLE 5.1   Dry Deposition of Atmospheric and Gaseous Sulphur and Nitrogen Compounds

Compound	Surface	Limiting Processes[a]	Mean Deposition Velocity (cm/s)	Range
SO_2 (gas)	Water	A	0.4	0.1−0.8
	Acid soil	S	0.3	0.1−0.5
	Alkaline soil	A	0.6	0.3−0.5
	Short vegetation, 0.1 m	S/A	0.5	0.1−0.7
	Medium vegetation, 1 m	S	0.8	0.1−1.5
	Tall vegetation, 10 m	S	0.5	0.1−2.0
HNO_3 (gas)	Water	A	0.4	0.1−0.8
	Most soils	A	0.8	−
	Most vegetation	A	1.0	−
NO_2 (gas)	Water	A	0.4	−
	Most soils	S	0.5	−
	Most vegetation	S	0.4	−
SO_4^{2-} NO_3^- (particles)	All surfaces	A	0.1	−

[a]A indicates atmospheric processes, and S indicates surface processes.

SOURCE: Fowler (1980).

can be much longer. Calculated residence times from all the currently conceived models are uncertain by a factor of at least two or three.

Even different species of a given element may have different residence times. Fogg and Fitzgerald (1979) found a residence time of 32 days for mercury removed from the atmosphere by rain. Species other than elemental mercury were thought to be involved, inasmuch as the equilibrium concentrations for mercury in water calculated from measured elemental mercury concentrations in air were much lower than the concentrations actually observed. Previously, Matsunaga and Goto (1976) suggested a residence time of 5.7 years, based upon measurements of mercury in air and precipitation. NRC (1978c) gives a much shorter estimate of residence time, 11 days. The discrepancy among these three residence times remains to be explained.

A number of atmospheric substances of different types and their residence times are given in Tables 5.2 and 5.3, and Figure 5.7. It must be borne in mind that values are only approximate and reflect the nature of the atmosphere where experiments were conducted (for example humidity, particle content, temperature, and altitude) as well as the experimental design.

The above physical and chemical characteristics of pollutants, which affect their residence time in the atmosphere, in turn affect their distribution in the biosphere relative to sources. For example, because of the regional to continental distribution of SO_x and NO_x, acid rain is very widespread (Figure 5.8), while the trace metals that are associated with particulate matter tend to be concentrated closer to sources (Figures 5.9 and 5.10). The short residence times of trace metals result in their being deposited in higher concentration close to urban areas, and of particular concern is the fact that many are present in precipitation at toxic concentrations (Figure 5.11).

Monitoring and Data Needs

While evidence is fragmentary, it is clear that a large number of toxic substances are emitted to the atmosphere and transported long distances, only to be redeposited in quantitites large enough to justify some concern. To assess and predict present and future effects of emissions upon ecosystems, we must improve our measurements of both atmospheric deposition and its ecological effects. We need not only a more adequate, long-term network of stations to measure atmospheric deposition (cf. Gibson 1979) but also a greatly improved understanding of the mechanisms by which materials are emitted to the atmosphere, transported from point sources over long distances by local winds and air-mass movements, and deposited in specific ecosystems. At present we know little about pollutants and their changes over time in cloud and rain drops (Scott 1978) or about the influence of various particulates upon the chemistry of precipitation. How gases and particles are deposited upon and absorbed by plant foliage also requires further study.

TABLE 5.2   Residence Times of Metals in the Atmosphere at
La Jolla and Ensenada.

| Metals | Days[a] | |
	La Jolla	Ensenada
Pb	7	8
Cd	0.7	0.5
Ag	0.2	0.1
Zn	0.4	0.3
Cu	0.5	1
Ni	3	0.8
Co	1.2	0.2
Fe	1.0	0.4
Mn	0.8	0.2
Cr	0.8	0.4
V	–	0.6
Al	1.0	0.2
Pb 210	5	–
Pu 239 + 240	1	–

[a]Standing crop of metals on particulates in a one kilometer high
square centimeter column of air divided by the flux, to a
square centimeter of ground surface. See reference for details
of measurement procedure.

SOURCE: Hodge et al. (1978).

TABLE 5.3   Average Residence Times in the Atmosphere of Substances Not Given in Table 5.2

Substance	Residence Times	Reference
O_3	0.4-90 days	Chatfield and Harrison (1977)
NO	4-5 days	Schlesinger (1979)
NO_2	2-8 days	Söderlund and Svensson (1976)
NO_3^-	4-20 days	Söderlund and Svensson (1976)
NH_4^+	7-19 days	Söderlund and Svensson (1976)
H_2S	0.08-2 days	Schlesinger (1976)
SO_2	0.01-7 days	Schlesinger (1976)
SO_4	3-5 days	Rodhe (1978)
Hg	11-2,080 days	Fogg and Fitzgerald (1979), Matsunaga and Goto (1976), NRC (1978c)
CH_3I	1 day	NRC (1976)
CO	0.9-2.7 years	Schlesinger (1979)
CCl_4	1 year	NRC (1976)
CH_4	1.5-2 years	Schlesinger (1979)
freon	16 years	NRC (1976)
CO_2	2-10 years	Schlesinger (1979)

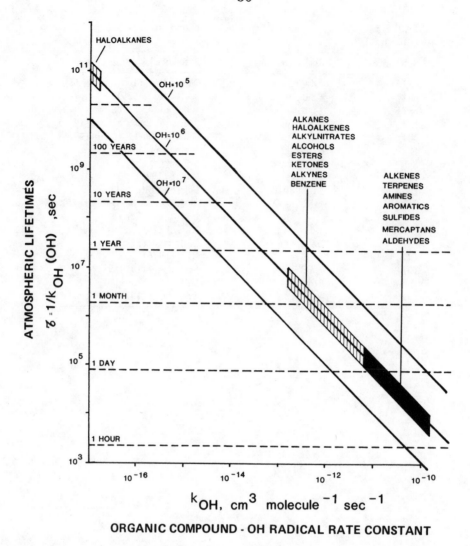

FIGURE 5.7   Atmospheric lifetimes of gaseous organic compounds. SOURCE: D. Grosjean, Environmental Research and Technology, Westlake Village, California, personal communication, 1980.

81

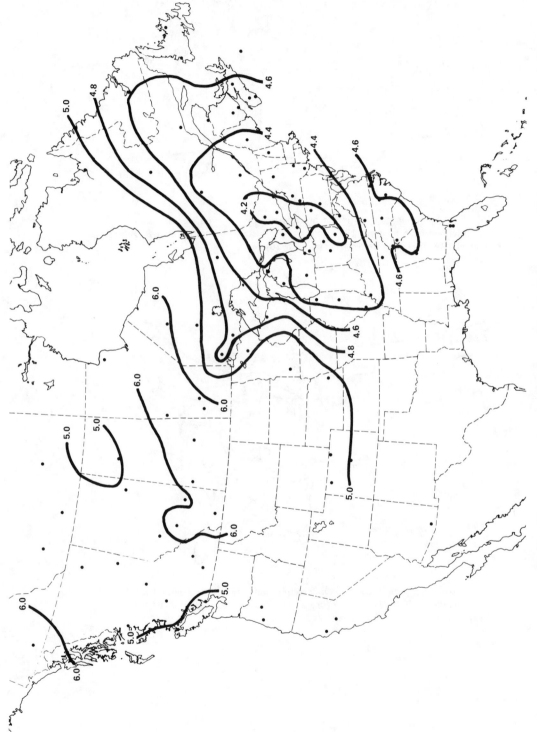

FIGURE 5.8 Mean annual pH of precipitation in the United States and Canada, 1979-1980. The isopleths are based on data from the National Atmospheric Deposition Program (NADP) in the United States and the Canadian Network for Sampling Precipitation (CANSAP) collected from April 1979 through March 1980. NADP data points are averaged from 16 or more weekly samples, and CANSAP sites of 4 or more monthly samples were included. SOURCE: W. W. Knapp, Department of Agronomy, Cornell University, Ithaca, New York, personal communication, 1981.

FIGURE 5.9 Average lead deposition by precipitation over the continental United States, September 1966 to March 1967 (in grams per hectare per month). SOURCE: Galloway et al. (1981); data from Lazrus et al. (1970).

FIGURE 5.10 Average zinc deposition by precipitation over the continental United States, September 1966 to March 1967 (in grams per hectare per month). SOURCE: Galloway et al. (1981); data from Lazrus et al. (1970).

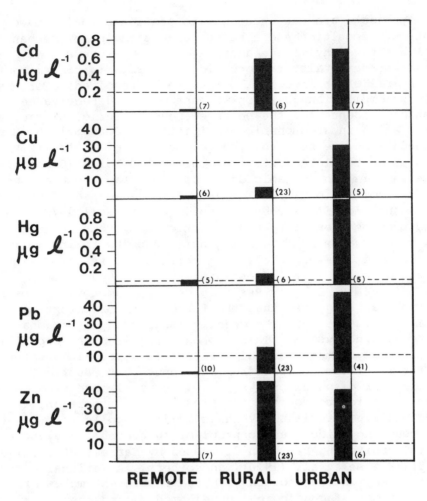

FIGURE 5.11 Median concentrations of metals in precipitation in remote, rural, and urban areas relative to organism toxicity levels. Each median in the figure is based on the number of data values designated in parentheses. Dashed lines denote threshold of organism toxicity reported by Gough et al. (1979). SOURCE: Galloway et al. (1981).

Several projects are now under way to provide both improved measurements of deposition and a better understanding of mechanisms (Table 5.4). Noteworthy is the absence of any measurement of deposition of trace metals, radionuclides, or organic contaminants. There are several other important points to be resolved. For example, on the acid precipitation question we must obtain direct rather than circumstantial evidence for the balance among sulfuric, nitric, and hydrochloric acids in atmospheric precipitation (Gorham 1976, Marsh 1978). In addition, we need to discover sources of neutralizing agents such as particulate dusts from cultivated soils (Gorham 1976) and gaseous ammonia from organic decomposition (Lau and Charlson 1977) or fossil fuel combustion (Gorham 1976). Ammonia can also be a strong acidifying agent once it reaches the soil (cf. Russell 1973, Reuss 1975b), and perhaps when it reaches aquatic ecosystems as well. The availability to organisms of both the nutrients and toxins in atmospheric deposition--especially the particulate fraction--is another phenomenon requiring investigation.

R.C. Harriss (NASA Langley Research Center, Hampton, VA; personal communication) has suggested that the forecasting of potential pollution in the future would be greatly assisted by a National Materials Accounting System. This approach requires the capability to monitor mass flows of materials of interest at appropriate points, including the materials' output from and input to production, transformation, or disposal processes. Industrial countries currently use such an operation to monitor the production, dispersion, and fate of many radioactive materials (Avenhaus 1977). Application of material accounting principles to problems in geochemistry and environmental monitoring has been discussed by Garrels et al. (1975), Ayres (1978), Kneese et al. (1970), and others. A national materials accounting system, in conjunction with reasonable estimates of future economic conditions, technological developments, and regulatory initiatives, would provide us with much improved predictions of future emissions of toxic substances to the atmosphere and their subsequent deposition.

SUMMARY

Transport, diffusion, and deposition processes affecting atmospheric substances are reasonably well understood, at least qualitatively. Knowledge of chemical transformations, particularly in the case of organic substances, is less complete. Despite the degree of understanding we have of processes, it is difficult to quantify the complete atmospheric pathway for a particular substance between source and receptor. Atmospheric residence time is a useful gross parameter by which to characterize the atmospheric behavior of a substance and its scale of influence. Knowledge of the latter as well as the geographical distribution of sensitive receptors is necessary for predicting the effects of a pollutant. The deposition of substances is a function of their physical state, particle size, reactivity,

TABLE 5.4 Major Studies of Atmospheric Transport and Deposition Currently Under Way in the United States

STATE	(Sulfur Transport and Transformation in the Environment) measures and models the transformation of SO_2 to SO_4 in point-source plumes and larger air masses and the subsequent transport and removal of these compounds. This program (U.S. EPA 1977, Wilson 1978) combines two earlier projects, MISTT (Midwest Interstate Sulfur Transformation and Transport) and the older RAPS (Regional Air Pollution Study) in the St. Louis area. Funded by U.S. EPA.
MAP3S	(Multistate Atmospheric Power Production Pollution Study) aims to improve ability to simulate changes in pollutant concentration, atmospheric behavior, and precipitation chemistry caused by changes in air pollution from large-scale, coal-fired power generation (MacCracken 1978, Mosaic 1979). Funded by DOE.
SURE	(Sulfate Regional Experiment) endeavors to predict ambient SO_4 levels in the atmosphere from SO_2 emissions by local sources (Perhac 1978, Mosaic 1979). Funded by EPRI.
APEX	(Atmospheric Precipitation Experiment) uses aircraft flights from the central United States out over the Atlantic Ocean, together with ground collectors, to examine the physics and chemistry of the processes by which airborne pollutants are deposited (Mosaic 1979). Funded by NSF and EPA.
NADP	(National Atmospheric Deposition Program) maintains 38 sites across the United States for weekly collection and analysis of major ions in wet and dry deposition (Mosaic 1979, Gibson 1979). Funded by USDA, USFS, USGS, EPA, NOAA, and DOE. The TVA also maintains 49 collectors for wet deposition within its area, and the World Meteorological Organization has 17 such stations in the United States, funded by EPA and NOAA.

solubility, and hygroscopicity, and of the surface characteristics of receptor organisms or surfaces.

Although this chapter has emphasized atmospheric processes and the behavior of pollutants, it is important to stress the role of the atmosphere as a reservoir of both beneficial and harmful substances. As one of the major portions of the biogeochemical cycle of many substances, and certainly for all of interest in this study, the atmosphere is the recipient of emissions from both nature and man, and it is also the source of these substances being delivered back into various ecosystems. The atmosphere and the biosphere are thus inseparably linked by the biogeochemical cycles of substances.

6. BIOLOGICAL ACCUMULATION AND EFFECT OF ATMOSPHERIC CONTAMINANTS

As we saw in the previous chapter, the particular structure of aquatic or terrestrial ecosystems can impose constraints upon interactions between the atmosphere and the biosphere. Terrestrial ecosystems are readily accessible to gaseous and suspended substances in the atmosphere. They present large, rough surface areas for exposure to gases, and where forest canopies impede wind transport, the wind velocity is reduced sufficiently to induce substantial dry deposition of fine particles. Evergreen canopies can be active in this manner year-round, whereas deciduous canopies are less effective in winter.

In contrast, aquatic ecosystems are somewhat protected from gases by the air-water interface. The rate of gas exchange between the atmosphere and a body of water depends upon the temperature and turbulence of the air and the water, and the diffusivity and concentration gradient of the gas across the air-water interface (Danckwerts 1970). The rate of exchange, however, is slow compared to aerial diffusion throughout terrestrial vegetation canopies.

ACCUMULATION IN TERRESTRIAL ECOSYSTEMS

In terrestrial ecosystems, atmospheric contaminants and pollutants are captured by plant surfaces in a variety of ways. Hairy and glandular plant surfaces can physically trap suspended matter in the form of wet or dry aerosols. Trace metals are an example of substances that tend to be tranported through the atmosphere and deposited in this way. Wet precipitation washes suspended matter from the air and also facilitates the eventual transfer of atmospheric contaminants captured by the vegetative canopy to the soil surface. Some metals, such as lead and cadmium, can even be incorporated as particles into plant cuticles (Hemphill and Rule 1975), and if the aerosol particles are less than 0.1 micron in diameter they can enter directly inside the leaf tissues through the stomatal pores. Ultimately all such material, whether attached to or incorporated into vegetation, is transferred into the soil surface by rainwater wash or by leaf fall and the decomposition of vegetation.

Aerosols may enter leaves through epidermal pores, trichomes, or wounds, as well as stomatal pores. In some cases direct penetration of the waxy cuticle is possible, or secretions by the plant may dissolve the aerosols, facilitating their transport and absorption by plant tissue (Hosker and Lindberg 1981). Factors that affect the size of stomatal apertures can determine how readily gases and aerosols reach the internal tissues of a leaf. The layer of moisture surrounding the palisade and mesophyll cells inside leaves also is crucial to the rate of reaction of the substances in the aerosols or gas with the living protoplasm. Among the substances that enter an ecosystem as aerosols are sulfur, nitrogen, and several of the more volatile heavy metals, such as mercury, selenium, and arsenic.

Gaseous pollutants generally react with plant surfaces more readily than aerosols. Some substances, such as hydrochloric acid, are so reactive that they will be absorbed by any plant surface they contact (W. Heck, USDA/SEA, North Carolina State University, Raleigh, NC, personal communication, 1980) while other gases, such as monoxides of carbon and nitrogen and volatile organic compounds, react very slowly with plants. Gaseous substances are also more readily incorporated into plant tissues and translocated by vascular systems than aerosol pollutants.

Most studies of pollutant uptake by plants have been of single pollutants on individual leaves or whole plants under laboratory conditions. Estimates of fluxes of pollutants into plant tissues under field conditions have been made using the eddy correlations technique (Eastman and Stedman 1977, Garland et al. 1973). At present there is no satisfactory way of extrapolating such data to whole ecosystems or to mixtures of pollutants. Figure 6.1 shows schematically some of the ways gaseous pollutants may react with plant surfaces, and Table 6.1 summarizes the types of reactions a number of gases have with plant tissues. Lindberg and McLaughlin (1981) have reviewed problems associated with collecting and interpreting data on the interaction of air pollutants with vegetation.

ACCUMULATION IN AQUATIC ECOSYSTEMS

The pathway followed by atmospheric contaminants entering surface waters depends on several factors, including thermal stratification in the water, the physical form of the contaminant at entry, its solubility, the interactions of organisms with the contaminant, and the affinity of the pollutant for sestonic (suspended) particles or sediment surfaces.

Aerosols and large particulate components of the atmosphere tend to become trapped in the surface film of natural waters. As a result, surface films often become enriched in atmospheric contaminants such as trace metals and organic micropollutants by as much as orders of magnitude more than the concentration observed in the bulk water beneath (Duce et al. 1972, MacIntyre 1974, Liss 1975, Andren et al.

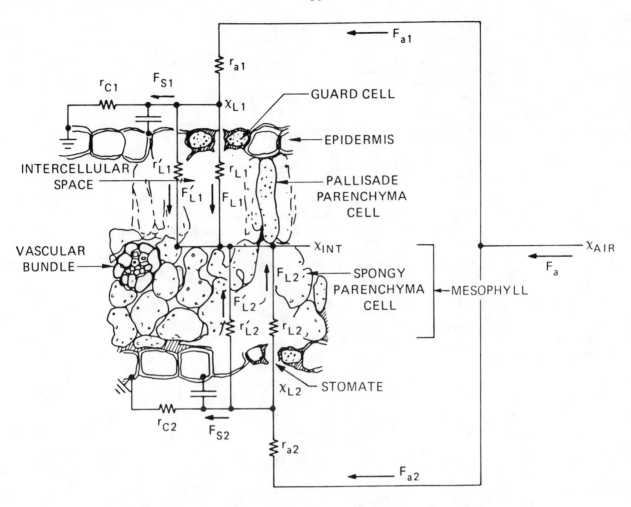

FIGURE 6.1 Electrical analog of pollutant exchange between leaf and surrounding air. The circuitry is superposed on a cross-section of an amphistomatous leaf (stomates on both surfaces). x_{air}, x_{L1} and x_{L2}, and x_{int} denote gaseous pollutant concentrations in well-mixed surrounding air, at the upper and lower leaf surfaces, and average gas phase concentrations within the leaf mesophyll, respectively. The (variable) resistances r_{a1} and r_{a2}, r_{L1} and r_{L2}, and r'_{L1} and r'_{L2} are the upper and lower boundary layer resistances, stomatal plus intercellular resistances, and cuticular plus internal resistances, respectively. Resistances r_{C1} and r_{C2} denote resistance to chemical reaction at the upper and lower leaf surfaces. The fluxes F_a, F_{a1} and F_{a2}, F_{S1} and F_{S2}, F_{L1} and F_{L2}, and F'_{L1} and F'_{L2} denote the total flux to both surfaces ($F_a = F_{a1} + F_{a2}$), surface uptake at the upper and lower surfaces, and the fluxes through the upper and lower stomata and cuticles. The electrical symbols depicting grounds ($\perp$) represent the termination of fluxes due to reaction with leaf surface materials. The (variable) capacitor symbols ($\dashv\vdash$) represent surface adsorption or retention (based on Bennett et al. (1973)). SOURCE: Hosker and Lindberg (1981).

TABLE 6.1   Types of Interactions of Gases with Plant Surfaces[a]

Type of Reaction	Atmospheric Contaminant
Primarily react with or are absorbed onto outer surfaces	H_2O_2, HNO_3, H_2SO_4
Primarily react within the leaf tissue	SO_2, NH_3, NO_2, PAN, metabolically reactive CO_2,[a] O_2[a] ethylene,[b] formaldehyde
React both on outer surfaces and in internal tissues	HF, HCl, O_3, Cl_2
React slowly but have potentially important interactions with plants	N_2O, NO, CO, organic molecules such as hydrocarbons, pheromones, and terpenes

[a]Normal large-scale exchange in photosynthesis and respiration.

[b]Ethylene is unusual in that it is a plant hormone and can have a strong effect in plant cells even at very low concentrations.

NOTE: Aerosols tend to react more with outer surfaces because they do not pass through stomata as easily as gases. Gases interact more readily with the internal tissues of plants because they can rapidly diffuse into the leaf through stomatal openings and the moist unprotected surfaces of internal tissues present more reaction sites for the gases.

SOURCE: After Hosker and Lindbery (1981).

1976, Elzerman and Armstrong 1979). The fates of such contaminants in surface films are poorly known.

Thermal stratification, either in the form of a thermocline or as ice cover, is a barrier to gas exchange and the dispersal of pollutants into aquatic ecosystems. Ice cover totally inhibits exchange of gases between air and water (Schindler 1971) and may be of major significance in temperate lakes, which can be ice covered for as much as 6 months, and in arctic lakes, on which ice cover can last more than 11 months of the year. In winter, oxygen consumption in ice-covered, eutrophic aquatic ecosystems may cause the entire water body to become anoxic. An experiment by Welch and coworkers (1980), simulating a natural gas pipeline break, found that methane discharged into an arctic lake in winter was trapped in the lake under ice cover until spring and that oxygen in the lake was slowly consumed by methane-oxidizing bacteria. On the other hand, atmospheric pollutants that accumulate in the ice cover of a lake over several months are discharged into the lake as a large pulse during ice and snow melt. Figure 6.2 shows one such pulse caused by the acid precipitation that accumulated in winter snow.

A larger-scale example of thermal stratification is the thermocline that exists between the surface ocean and the deep ocean. Except at the poles where mixing can occur the thermocline acts as an effective barrier against transfer of substances between surface and deep ocean water. For example, the time needed to achieve equilibrium in CO_2 transfer across the ocean thermocline is several hundred years, while the time required for equilibrium between the atmosphere and the surface water of the ocean is only about 8 years. If the ocean surface waters had sufficient capacity for CO_2, the equilibration time is short enough that the CO_2 released by anthropogenic activities would not accumulate in the atmosphere. The surface ocean, however, cannot absorb all the CO_2 that is released. Exchange with the deep ocean is therefore the process that limits the rate of exchange and the achievement of equilibrium for CO_2 between the atmosphere and the oceans, and the exchange is so slow that excess CO_2 does indeed accumulate in the atmosphere. Likewise, once a substance enters the deep ocean, residence time is very long, not only for carbon dioxide but also for other contaminants emitted into the atmosphere as combustion products (Broecker 1974).

Carbon dioxide, sulfate, nitrate, and ammonium are all required to some degree by algae and other aquatic plants. The carbon, sulfur, and nitrogen are either actively metabolized and passed up the food chain or egested as fecal pellets. Although much of the fecal material is remineralized in the water column, some of the carbon, nitrogen, and sulfur eventually collect in sediments. Nitrogen, which is usually in greatest demand by plants, tends to be transferred most rapidly. Reactive pollutants adsorbed or absorbed by plankton or other suspended articles travel similar pathways. Less reactive pollutants, which remain in solution, tend to remain trapped above the thermocline, because diffusive transfer across the thermocline is much slower than the settling of particles.

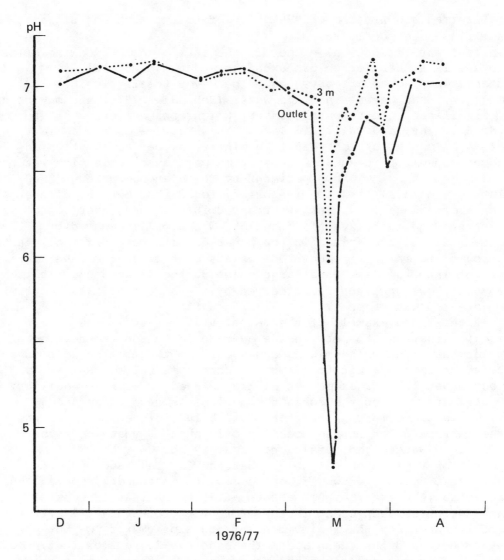

FIGURE 6.2 Seasonal changes of pH of lake water taken from the outlet of Little Moose Lake, Adirondacks, New York, and from a 3-m-deep pipe in the lake. The spring acid pulse is greater at the surface than at the 3-m-deep level. SOURCE: Schofield (1980). Reprinted with permission from Ann Arbor Science Publishers, Inc., Ann Arbor, Michigan.

For example, Hesslein et al. (1980) studied the pathways of trace metals by adding radioactive tracers to a whole lake. Iron-59 and cobalt-60, which are quickly adsorbed by particulate matter, settled rapidly from the epilimnion (surface water) to sediments. Iron redissolved in anoxic surface sediments underlying the hypolimnion (deep water). In contrast, cesium-134 remained in solution and was removed from the epilimnion at a much slower rate, chiefly by direct reaction with littoral sediments. Removal times were roughly twice as long as for cobalt and iron. Other metallic tracers had intermediate removal times.

Substances that are relatively inert can accumulate in the deep ocean. Examples are the radioactive pollutants argon-39 and krypton-85, or the halocarbon pollutants CF_3Cl, CF_4, and CCl_4. All of the above enter surface waters as gases. Cesium-137, which is a product of weapons testing and the nuclear fuel cycle, enters the ocean as an aerosol. It accumulates in the lipid fractions of marine organisms, and from the lipid phase it is ultimately released by metabolism or by rapid mineralization upon the death of the organism. Hence, cesium is removed from salt water as slowly as from fresh water.

The volume of the deep ocean is 50 times that of the layer above the thermocline. Pollutants entering the deep ocean are thus diluted 50-fold from their concentration in the surface layer. Some will enter the sediments associated with the depositing inorganic or biological debris. Those with longer residence times--i.e., the less reactive pollutants--will remain in the ocean deeps for millennia.

ACCUMULATION IN SOILS AND AQUATIC SEDIMENTS

In both terrestrial and aquatic communities, there is a tendency for most pollutants to accumulate in soils or lake sediments, where they become concentrated by absorption, direct cation exchange, or by chemical precipitation. There are both striking similarities and substantial differences in the response of soils and aquatic sediments to pollutants. Both are depositories for biological and mineral matter and can concentrate any pollutants bound to deposited materials. In terrestrial ecosystems, however, plant roots tend to resorb and recycle materials to some degree, and there is some long-term storage in wood. With the exception of sphagnum bogs and wetlands, there is no such recycling in aquatic ecosystems, and there are no long-lived plants to form long-term living sinks for pollutants although dead organic debris may accumulate in aquatic sediments that are long-term sinks for many substances in the form of organic molecules, insoluble oxides, hydroxides of redox-sensitive metals, or as sulfides under anoxic conditions.

In anoxic environments, there may be substantial recycling of many elements that are not mobile in oxic environments from sediments into the hydrosphere and atmosphere. For example, sulfur may be released as hydrogen sulfide or dimethyl sulfide; several metals are released in a methylated form; and nitrogen is released as ammonia, nitrous oxide, and molecular nitrogen. These processes, widespread in aquatic ecosystems, also occur in waterlogged soils (Tusneem and Patrick 1971, Reddy et al. 1978). Flux rates for the above substances are poorly known and are highly variable even within one ecosystem.

In aquatic ecosystems, some pollutants are rapidly locked into sediments, where they cease to affect planktonic or nektonic organisms. For example, the addition of phosphorus to lakes is objectionable largely because it causes increased growth of phytoplankton. Once deposited in oxic sediments, it is usually no longer a threat. Sediment-water partition coefficients for phosphorus range from 10^4 to 10^6. Likewise, most trace metals are rapidly transferred from water to sediments (Hesslein et al. 1980), as are nitrogen and sulfur, although they are less efficiently concentrated in sediments because microbial reduction can rapidly release them (Schindler et al. 1977, 1980a).

Under certain circumstances, the high sediment-to-water partition coefficient may become an undesirable feature for the aquatic ecosystem as a whole. Once incorporated into sediments, a pollutant cannot be flushed rapidly from a water body. Some materials may be transferred directly from sediments to pelagic biota; for example, mercury, selenium, and perhaps lead are methylated in sediments and transferred to the water column. Owing to their high affinity for lipids, methylated metals can be taken up and accumulate in fish and other pelagic forms, even though concentrations in water remain low. DDT and many other pesticides, petroleum hydrocarbons, polynuclear aromatic compounds, and polybrominated or polychlorinated biphenyls (PBBs and PCBs) also are known to be highly concentrated in both sediments and the lipids of living organisms.

Bottom-living aquatic organisms that ingest sediments many receive very high concentrations of pollutants. The incidence of deformities among the larvae of bottom dwelling chironomids is one of the most sensitive indicators of contamination in the aquatic environment (Warwick 1980a, 1980b). Likewise, surface soils, where high accumulation of pollutants occurs, are occupied by innumerable small invertebrates, including mites, springtails, beetles, earthworms, Crustacea, and midge larvae, and by a variety of fungi, bacteria, and other organisms. The fungi include both important plant pathogens and mycorrhizal fungi crucial for recycling nutrients from soil to plants. All these organisms perform decomposition via many complex pathways, recycling crucial nutrients in forms usable by plants. The seeds, spores, and eggs of many organisms also begin development in these layers.

Saprophytic organisms that burrow throughout soil and sediments may cause deposited pollutants to be diluted, or spread throughout the upper layers. As a result, substances deposited in one year may be dispersed through a layer of soil or sediment representing decades or even centuries of accumulation (Davis 1968, Berger and Heath 1968, Rhoads 1973). Such processes will delay both the development of high concentrations of pollutants in surface sediments and the burial of substances after inputs have ceased (Schindler et al. 1977). This "bioturbation" may be responsible for the fact that concentrations of DDT and PCBs in the fishes and sediments of the Great Lakes have not decreased greatly in the past decade, despite stringent control of inputs (International Joint Commission 1980).

Acid conditions and high trace-metal concentrations in excess of 500 ppm are known to inhibit litter decomposition, causing undecomposed organic matter to accumulate in both terrestrial and aquatic ecosystems (Watson et al. 1976, Freedman and Hutchinson 1980, Strojan 1980, Hendrey et al. 1976, Traaen 1980). A number of key enzyme systems appear to be affected (Tyler 1974, 1976a,b). This accumulation assuredly slows nutrient cycling.

Soil and aquatic fungi are usually more acid tolerant than bacteria, and thus they become the dominant decomposers in acidic ecosystems. But even fungi appear to be susceptible to extremely high concentrations of metals derived from the atmosphere (Jordan and Lechevalier 1975, Williams et al. 1977). Fungi with increased tolerance of acids and heavy metals have, however, been obtained from several sites near smelters (Hartman 1976, Freedman 1978, Carter 1978).

Aquatic Sediments as Historical Records

Aquatic sediments provide a "history" of changes in the aquatic ecosystem, its terrestrial watershed, and its airshed. The increase in heavy metals in lake sediments in this century is an excellent example (Robbins and Edgington 1977, Norton et al. 1978, Galloway and Likens 1979). A study of sediment cores from the mouths of tributary streams around the St. Lawrence River and Great Lakes confirms the relationship between such increases and terrestrial watersheds contaminated by human activity (Fitchko and Hutchinson 1975). Mackereth (1966) and Warwick (1980) were able to show historical evidence for increased weathering of terrestrial watersheds owing to human disturbance. Davis et al. (1980) have been able to decipher an acidification history of lakes in Norway and Maine, using a combination of recent dating methods and changes in the abundance of acidophilic diatoms. No study has yet used paleolimnological methods to construct a history of terrestrial responses to acidification. In most cases, such methods offer at present the only possibility of documenting and dating the course of acidification in either terrestrial or aquatic ecosystems, because of the paucity of regional background data collected prior to the development of the acid rain problem.

TRANSFORMATIONS IN SOILS AND WATERS

A number of volatile components of the atmosphere can react with waters or soil constituents and be incorporated into the compounds in the biosphere. How volatile components are removed, the responsible organisms or abiotic mechanisms, and frequently the factors that can affect removal are as yet unclear or wholly unknown. Some removal mechanisms are abiotic, but more often they are entirely biological.

Many organic compounds present in the atmosphere undoubtedly are transformed and decomposed in soils and waters. The evidence for such transformations, however, is derived usually from studies in which the chemical has not been supplied in the volatile form and in which the chemicals were present in water or soil samples at concentrations far higher than are present in nature. Tests with concentrations of contaminants far in excess of ambient levels must be interpreted very cautiously, because chemicals that are biodegraded at high concentrations may degrade slowly or even not at all when present at low concentrations (Boethling and Alexander 1979a,b). If the transformation yields inorganic products (CO_2 and H_2O), the reaction detoxifies the molecules.

Among the organic compounds destroyed--at least at higher concentrations--are certain polycylic aromatic hydrocarbons, phenols, alkanes, esters, and simple aromatic molecules. In some instances, however, the transformations may yield not inorganic products but other organic products, which may themselves be toxic. Certain organic compounds are resistant to microbial transformation and thus persist, for example certain chlorinated aromatics, aliphatics, and several nitrogen heterocycles. The biodegradability of many of the organic atmospheric pollutants has not been directly evaluated, and none has been tested at realistic concentrations. Hence it is not possible at the present time to assess their rate of transformation when introduced into waters or soils or even whether such a transformation does take place. Some of the problems in assessing such kinds of transformations and the types of reactions that occur are considered by Alexander (1981).

Information on the role of soils in removing or generating atmospheric substances is growing constantly, and current assessments will need to be modified as information is obtained from new studies. The following examples have been relatively well researched and are offered as an indication of the current level of understanding for several atmospheric constituents.

Methane. A number of bacteria present in soils and lake waters are able to oxidize methane. Although there has been considerable study of which organisms can do so in laboratory media and of the biochemistry of the process, surprisingly little attention has been paid to the rates of biological methane oxidation under natural conditions. Studies of the process in paddy fields indicate that the rate of oxidation is limited by the rate of methane diffusion to the soil (de Bont et al. 1978). In lakes, the rate of oxidation is usually limited by diffusion of oxygen and methane, as well as by the concentration of nitrogen (Rudd et al. 1976). During summer thermal stratification, activity is restricted to the metalimnion (the part of the water column where the thermal gradient occurs), although after the fall overturn destroys thermal stratification methane oxidation may occur throughout the water column. Often this activity persists throughout the following winter (Rudd and Hamilton 1975, Welch et al. 1980).

Carbon monoxide. An increase in ambient levels of carbon monoxide would be expected on the basis of present rates of global emission, but no such major increases are evident; therefore a significant sink for this gas must exist. Many investigators have pointed to soils as a means for removing carbon monoxide from the gas phase. Because carbon monoxide is oxidized in nonsterile but not in sterile soil (K. Smith et al. 1973), the removal mechanism is apparently microbial. Estimates have been made of global uptake by soil of carbon monoxide based on its rapid loss from the head space of a container when soils are incubated in the laboratory with this volatile compound.

For example, based upon tests in which several soils were exposed to 100 ppm carbon monoxide, it was calculated that soils of the world remove 14.3×10^{15} g of carbon monoxide per year (Ingersoll et al. 1974). Such figures are undoubtedly a gross overestimate, because the rate of removal is markedly dependent upon concentration and appears to follow Michaelis-Menten kinetics. The rate of removal at concentrations equivalent to those in the atmosphere is far lower than when abnormally high concentrations are used in the tests (G. W. Bartholomew and M. Alexander, Cornell University, personal communication). Other estimates suggest that the earth's land surface removes 8×10^{9} g of carbon monoxide per hour (Heichel 1973) and that the rate of removal ranges from 4.5 to 14×10^{14} g per year (Nozhevnikova and Yurganov 1978). Despite the disparity in estimates for removal, it is generally believed that soils are a major sink and that microorganisms in soil are chiefly responsible for the removal (NRC 1977d, Liebl and Seiler 1976).

Many individual microbial species are able to metabolize carbon monoxide in laboratory culture (Nozhevnikova and Yurganov 1978, Bartholomew and Alexander 1979). Investigations with soils suggest that, in the few soil samples investigated, the microorganisms consuming carbon monoxide are neither heterotrophs using carbon monoxide as a carbon source nor autotrophs using it as a source of energy; rather the responsible organisms co-metabolize the gas (Bartholomew and Alexander 1979, and personal communication).

Ethylene. Soils also remove ethylene from the gas phase. Little of this removal occurs in sterilized or air-dried soil, but the reaction is rapid in moist and nonsterile soil. Such data indicate that the process is microbial. The organisms involved are apparently aerobic species, probably bacteria. The few rates that have been calculated in laboratory studies indicate that from 0.14 to about 14 nmol can be removed per gram of soil per day (K. Smith et al. 1973, Abeles et al. 1971). In aquatic ecosystems ethylene is oxidized by the same bacteria that oxidize methane. Under similar conditions ethylene usually is used preferentially (Flett et al. 1975).

Acetylene. Acetylene can apparently be readily destroyed, again by microbial processes, since it does not occur in sterilized or air-dried soil. The rate of removal of acetylene at a moisture level of 50 percent of the water-holding capacity of the soil is 0.24 to 3.12 nmol per gram of soil per day, depending on the composition of the soil (K. Smith et al. 1973). Acetylene can be reduced to ethylene by nitrogen-fixing bacteria in both aquatic and terrestrial ecosystems--a fact that is utilized to estimate amounts of nitrogen fixation (Stewart et al. 1967).

Nitrogen. Biological nitrogen fixation leads to N_2 incorporation into soil, and one global estimate for this process is that the annual amount of molecular nitrogen fixed into soil is 99 x 10^{12} grams of nitrogen per year (Delwiche and Likens 1977). Both blue-green algae and methane-oxidizing bacteria are known to fix N_2 in aquatic ecosystems (Davis et al. 1964). While there have been some measurements of nitrogen fixation in aquatic ecosystems (Johannes et al. 1972, Mague and Holm-Hansen 1975) data are presently too scarce to establish their contribution to global fixation of nitrogen.

Nitrous oxide may also be removed from the gas phase. This removal is also apparently related to microbial activity and is part of the sequence involving the reduction of inorganic nitrogen compounds to molecular nitrogen. The process is promoted by anaerobiosis and by the presence of organic materials that stimulate microbial proliferation. But soils are a more significant source than a sink for nitrous oxide (Blackmer and Bremner 1976).

Nitric oxide is sorbed from the air. This sorption may take place even in dry soil, and the process leads to an increase in acidity and a rise in the nitrate concentration as the sorbed gas is oxidized in soil to nitrate (Prather et al. 1973a).

Nitrogen dioxide is also readily removed from the gas phase, and it is converted in soil to nitrite and nitrate. Nitrogen dioxide is sorbed rapidly onto air-dried soils, and the removal is apparently nonmicrobial, because it takes place in both nonsterile and sterile soil (Prather et al 1973b). On the other hand, the nitrite that is formed is converted to nitrate--and is thus detoxified--by microorganisms (Ghiorse and Alexander 1976). Soils remove a smaller portion of ammonia by chemical reaction because much of the ammonia is removed by rainfall and dry deposition (Rodgers 1978).

Sulfur. Sulfur dioxide is readily removed from the atmosphere in contact with soil. The removal appears to be non-biological because rates are the same whether the soil is sterile or nonsterile. Much of the sulfur dioxide that is thus removed by soil is converted to inorganic sulfur products, and much of it is probably ultimately converted to sulfate. Nevertheless, part of the sulfur derived from the gas phase is apparently converted to an organic sulfur compound or compounds (Ghiorse and Alexander 1976). The removal is affected by the moisture content of the air and the sulfur dioxide concentration of the gas phase (Yee et al. 1975).

Soils also remove hydrogen sulfide and methyl mercaptan from the air. The addition of water to a dry soil decreases the rate of methyl mercaptan removal and has little effect on hydrogen sulfide removal. Both air-dried and wet soils are able to remove dimethyl disulfide, carbonyl sulfide, and carbon disulfide. In the process, small amounts of carbonyl sulfide are converted in moist soils to carbon disulfide.

Microorganisms are apparently involved in the removal of dimethyl sulfide, dimethyl disulfide, carbonyl sulfide, and carbon disulfide by moist soils, but they may not be significant in the removal of hydrogen sulfide, which reacts chemically (K. Smith et al. 1973, Bremner and Banwart 1976). The ultimate destruction of the organic portions of any sulfur compounds that are generated probably requires microbial intervention, as is true of most organic compounds.

TRANSFER OF SUBSTANCES FROM TERRESTRIAL TO AQUATIC ECOSYSTEMS

Because aquatic ecosystems are repositories not only for direct atmospheric deposition but also for many of the substances leached from or initially deposited upon terrestrial drainages, aquatic ecosystems may be contaminated to a greater degree than terrestrial ecosystems. The degree to which materials entering from the atmosphere are retained by terrestrial soils and vegetation is therefore an important consideration when aquatic contamination is studied, particularly because the terrestrial portion of a watershed is usually much larger than the lake or stream which drains it.

Many substances have a high affinity for clay minerals in soils and become tightly bound in terrestrial soils. Large amounts of such substances will not be transferred to lakes as long as soil erosion is well controlled. This includes most trace metals and the hydrogen ion, as well as phosphorus. Other substances, such as nitrogen, are retained due to biological demands. At Hubbard Brook, New Hampshire, over a 10-year period, 88 percent of the ammonium that entered the watershed was retained, but only 13 percent of the nitrate was retained (Likens et al. 1977).

Still other substances leave watersheds in amounts equal to or higher than that in the precipitation that enters the watershed. At Hubbard Brook, stream flow removed more sulfate than entered as bulk precipitation. The discrepancy may be due to dry deposition, which was not included in estimates of substances entering the system. In Norway, inflow and outflow of sulfate from lakes are roughly balanced (Abrahamsen 1980). The difficulty in measuring entry of a substance to a watershed via dry deposition makes more precise estimates of watershed balances impossible. Many cations may leave terrestrial soils after exchanging for H+. These include both soil nutrients, such as calcium, and substances which may be toxic to aquatic life, such as aluminum.

One study of substances deposited from the atmosphere to terrestrial soils (Hutchinson et al. 1975) showed that nickel, copper, and cobalt from the Sudbury smelter had not only accumulated in soils to concentrations toxic for seedlings but had also poisoned nearby lakes through runoff from the watersheds. River sediments in drainage from the area were contaminated up to 100 km away. Land management practices that allow land erosion or leaching are thus likely to enhance the movement of atmospheric substances into aquatic ecosystems.

It now is recognized that fish mortality attributed to acid precipitation is in large part caused by the aluminum leached from

terrestrial soils during spring melt (Cronan and Schofield 1979).
Acid precipitation is also identified as a contributor to the
increased runoff of nitrate, which is causing concentrations of
nitrate in Adirondack surface waters to become uncomfortably close to
maximum acceptable levels for drinking water. While much of the
nitrate undoubtedly results from precipitation inputs in excess of
biological demands, it is possible that the suppression of
denitrification by acidification of soils may favor nitrate
accumulation (Alexander 1980).

Seip (1980) calculated that the actual input of hydrogen ions to
Norwegian lakes from direct precipitation as well as its runoff from
the lake catchment basin, could explain only a negligible part of the
acidification that takes place. The rate of acidification is better
correlated with sulfate deposition. Sulfate output from terrestrial
ecosystems is roughly equal to input (Abrahamsen 1980), and thus if
Seip's observations are correct, increased sulfur deposition in
terrestrial areas could contribute to the acidification of receiving
waters. A complete understanding of the natural processes and
mechanisms involved in the acidification of freshwaters is still
needed.

On the other hand, few lakes are as acid as the precipitation
falling in their watersheds, which leads some authors to believe that
buffering by terrestrial ecosystems may protect poorly buffered lakes
from acidifying as rapidly as they might under the influence of
unaltered acid precipitation upon the entire drainage basin (Gorham
and McFee 1980). The relatively high pH levels of calcareous soils
neutralize the acid, while in soils below pH 5, aluminum species may
be the major source of buffering (Johannessen 1980).

The magnitude of transfers of acidifying substances from
terrestrial ecosystems to aquatic ones is usually unknown, and this
hinders the prediction of rates of acidification and recovery of
aquatic ecosystems. Henriksen (1979) and Almer et al. (1978) have
developed simple models, based on expected ratios of alkalinity to
concentrations of calcium or calcium plus magnesium for predicting the
degree of acidification of lakes. Henriksen (1980) has extended this
logic to predict the degree of acidification which will take place
under precipitation regimes of different acidity. The model assumes
that as precipitation becomes more acid, the calcium concentration of
lakes does not increase. Several studies support this assumption.
Malmer (1974) and Schofield (cited in Henriksen 1979) found no
increase in calcium concentrations of waters in acidified areas.
Schindler et al. (1980a) found no increase in calcium release from
lake sediments as pH decreased. On the other hand, Gordon and Gorham
(1963), Almer et al. (1974) and Dillon et al. (1980) have observed
that calcium losses are higher from terrestrial ecosystems in areas
where the pH of precipitation is from 4.0 to 4.5 than in areas with
precipitation of normal acidity. It is probable that these
observations reflect differences among soil types. Abrahamsen's
(1980) experimental acidification of forest plots in Norway supports
the latter argument. If the relationship between calcium leaching and
acid inputs can be quantified, Henriksen's model may have predictive

value (Figure 6.3), although different soils may respond differently to acid inputs.

The surface film in aquatic ecosystems is an area of concern. As noted earlier, organic pollutants and heavy metals may concentrate there at orders of magnitude higher than in the atmosphere above or the bulk water beneath. The ecology of the biota in such layers (the neuston) is little known. However, it is known that a wide variety of organisms congregate near the surface including microscopic autotrophs (algae) and heterotrophs (e.g., amoebae, bacteria), as well as larger organisms such as pelagic fish eggs and crustacean larvae (Makarov 1976). Oil spills as well as atmospheric deposition may add toxic organic molecules to the surface film and thus affect the survival of the inhabitants of this unique community. For example, recent studies by Hardy and Creselius (1981) have shown urban aerosols to be six times more toxic than rural particles, with toxicity to marine phytoplankton being due primarily to soluble Pb, V, Cd, Cu, Zn, and Ni, in that order. While present deposition rates of atmospheric particulate matter do not appear sufficient to inhibit marine primary productivity, they may have serious effects on the sea-surface microlayer--the neuston.

Wetlands are transitional ecosystems that incorporate many of the features of both terrestrial and aquatic ecosystems, such as canopy capture and air-water exchanges. In addition, gaseous substances may be regenerated by both oxic and anoxic means. There are vast tracts of wetlands in northern latitudes and coastal marine areas, but data for such ecosystems are too scarce. This shortcoming must be rectified to provide an accurate evaluation of their role in global biogeochemical cycles, and to permit natural global exchanges between the atmosphere and the biosphere to be quantified.

POLLUTANTS AND INDIVIDUAL ORGANISMS

Several possible types of interactions may occur between pollutants and individual organisms. The pollutant may be actively taken up by the organism--bioconcentrated--or, avoided or excreted--bioexcluded; neither occurs if the molecule is biologically inert. In general, substances required for the growth of the organism will be bioconcentrated. For example, aquatic plants require nitrogen in much higher concentrations than exist in surrounding waters, and they may concentrate the element up to 30,000 times with respect to the water in which they grow (Vallentyne 1974). Phosphorus and carbon, essential nutrients to all living forms, are also bioconcentrated by plants.

Some other elements, nutrients not essential to most organisms, are bioconcentrated by organisms with special requirements. For example, only diatoms and a few other Chrysophyta will concentrate silicon, which they require for their frustules. Molluscs and vertebrates require much more calcium per unit weight than organisms that have no heavily calcified structures in their bodies.

102

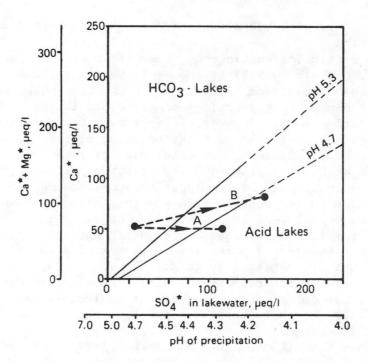

FIGURE 6.3 Nomogram used to predict changes in the acidity of lakes as the acidity of precipitation changes. If changes in the acidity of precipitation cause no increase in the weathering of the basic cations, Ca^{++} and Mg^{++}, the relationship between the concentration of basic cations and the concentration of sulfate in lake water would be expected to move horizontally as a lake is acidified (arrow A). If increased acidity is accompanied by increased leaching of basic cations, the lake water chemistry changes as indicated by the arrow B, causing the degree of acidification to be underestimated. SOURCE: After Henriksen (1980).

Some rare elements are bioconcentrated because an organism's physiology cannot distinguish them from more abundant and useful elements that are chemically similar. The fact that molluscs and vertebrates take up strontium, lead, and radium along with calcium is well-known. Incorporation of toxic or radioactive substances with calcium is particularly dangerous, because substances incorporated in bone, shells, and other calcified tissues are not metabolized quickly--i.e., they have a long biological residence time. In general, elements with similar chemical properties may be expected to be dealt with in a similar fashion by organisms.

A special form of bioconcentration, called biomagnification, occurs when organisms bioconcentrate substances via the food chain. The accumulation and concentration of DDT in predatory birds is a well-known example of the biomagnification of this toxic substance.

More commonly, toxic elements are bioexcluded, that is, organisms contain them at concentrations much lower than those in their surroundings or their food. For example, vertebrates appear to exclude lead selectively from absorption by their bodies; the concentration of lead in their tissue is lower than in food organisms (Settle and Patterson 1980). Gächter and Geiger (1979) using radioactive tracers found that 5 toxic metals, including mercury and zinc, decreased in concentration as they were passed up an aquatic food chain. Their results contrast with other published reports that show mercury to be bioconcentrated (National Research Council Canada 1979). For all metals, phytoplankton and periphyton contained the highest concentrations, zooplankton and insects somewhat less, and fishes least of all.

Microorganisms can make methylated derivatives of mercury, selenium, and other trace metals (Wood 1974, Chan et al. 1976, Braman and Forbach 1973). The decomposers can then bioexclude these trace metals owing to the solubility and volatility of their derivatives, which are rapidly carried away by the water column. On the other hand, methylated metals are known to be bioconcentrated in pelagic vertebrates and other organisms where they accumulate in lipid deposits (Wood 1974, Benson and Summons 1981).

In between bioconcentration and bioexclusion are biointermediate elements, which are concentrated only slightly more in organisms than in water, and biologically inert substances, which are found at similar concentrations in both the organism and its surroundings. Oxygen and chloride are nearly biologically inert; potassium and sulfur tend to be biointermediate.

Bioconcentrated substances may be stimulatory or very toxic. Nitrogen and phosphorus concentrated by plants from their surroundings stimulate plant growth, if some other factor is not limiting. On the other hand, high concentrations of arsenic are known to inhibit the uptake of phosphorus by phytoplankton slowing the growth of some organisms (Brunskill et al. 1980). In some cases, a substance may stimulate one group of organisms while being toxic to another.

When two or more pollutants enter an ecosystem together, one may enhance or lessen the effects of the other. For example, brook trout are more susceptible to aluminum poisoning at high concentrations of hydrogen ions, owing to the effects of the latter on the chemical form of aluminum in solution (Cronan and Schofield 1979). On the other hand, the toxic effects of mercury on a wide variety of organisms are suppressed by high concentrations of selenium, through the latter is itself a toxic substance (Rudd et al. 1980).

The same nonmetabolic absorption-adsorption reactions that occur in clays and inert particles take place on the outer surface of some aquatic organisms. The silicious outer frustules of diatoms are known to concentrate radium by adsorption (Havlik 1971, Edgington et al. 1970, Emerson and Hesslein 1973).

Physical factors, such as temperature and light, directly and indirectly affect biological responses to pollutants. Phytoplankton depend on light for photosynthesis and growth, which means that the uptake of any element by plankton is ultimately dependent on light. All of the physiological parameters in biota are temperature dependent to some degree. The solubility of gases and the kinetics of chemical reactions are also dependent on temperature.

In some cases, near-lethal concentrations of substances occur in food chains, even under natural conditions. In older carnivorous fish, bioconcentration may lead to 1 to 2 milligrams of mercury per kilogram of flesh, even in remote areas of the Arctic. Under such circumstances, even small anthropogenic additions to the ecosystem can lead to concentrations exceeding standards for human consumption. Armstrong (1979) mentions that fish with concentrations of mercury greater than 10 mg/kg fresh body weight are seldom encountered even in the most heavily polluted environments and hypothesizes that such concentrations may be lethal to the fish themselves.

Toxicity of Deposited Substances to Organisms

A number of the substances deposited in ecosystems as a consequence of fossil fuel combustion are known to be extremely toxic to plants or animals. These include: SO_2, NO_2, SO_3, H_2SO_4, NO_2, HNO_3, HF, H_2S, and various trace organics and metals. In most situations, the amount deposited is lower than concentrations reported to have toxic effects, but there are some exceptions, and because few toxicity studies cover entire life cycles for sensitive test organisms or combinations of several pollutants, predictions should be very conservative.

Terrestrial Ecosystems

In general, primary gaseous pollutants affect only terrestrial organisms and ecosystems. The distance over which effects are noted depends, of course, on weather conditions, the height of the stack

emitting the gaseous pollutant into the atmosphere, and the
sensitivity of the receptor organisms.

Of most concern is sulfur dioxide, which affects sensitive species
at extremely low concentrations or after very short exposures. As
discussed later, SO_2 may also interact synergistically with other
gaseous pollutants--especially ozone--causing increased damage.
Linzon (1971) found that Pinus strobus, eastern white pine, was
damaged after several years' exposure to an average SO_2
concentration of 8×10^{-3} ppm. This concentration is approached in
many rural and wilderness areas of the United States and southern
Canada, for example in northern Minnesota and northwestern Ontario
(Glass and Loucks 1980) and the eastern United States (Costonis
1972). In other studies, fumigations of only one to several hours
with 2.5 to 5 pphm SO_2 caused visible damage to new white pine
needles (Costonis 1970, 1972, 1973; Houston 1974). Germination and
seed production are also reduced at low SO_2 concentrations,
particularly when other gaseous pollutants are present (Houston and
Dochinger 1977). Lichens and bryophytes have been described as
especially sensitive to SO_2. Many species of these plants are
absent in urban areas and around point sources where SO_2
concentrations for extended periods exceed 0.02 ppm SO_2 (Hawksworth
and Rose 1976). The effects of acid precipitation on terrestrial
ecosystems are covered in several recent publications (Hutchinson and
Havas 1980, Drablos and Tollan 1980, Overrein et al. 1980).

Although sulfur dioxide is the pollutant of primary concern with
vegetation, nitrogen oxides have been shown to cause some damage when
they are combined with sulfur dioxide, a typical situation near many
sources of emissions. Nitrogen oxides can also interact with
hydrocarbons to form secondary pollutants such as ozone and
peroxyacetylnitrate (PAN), which are more toxic than the nitrogen
oxides at low concentrations (Dvorak et al. 1978, Cleveland and
Graedel 1979). The Academy report on ozone and other photochemical
oxidants (NRC 1977b) provides some detailed data on damage by those
pollutants to a number of plant species, and another report (NRC
1978a) summarizes similar information for sulfr oxides. Responses of
sensitive vegetation to nitrogen oxides are given in Figure 6.4. The
developmental stage of a plant influences its sensitivity; many plants
appear to be most susceptible to damage from gases during flowering.

There are no definitive studies of the effects of gaseous
emissions on animals at concentrations observed in the environment,
although effects have been predicted. Animal populations in fumigated
ecosystems may be affected by the disappearance of sensitive plants
that are their usual food or shelter (Dvorak et al. 1978). Along
roadsides, for example, contamination from atmospheric pollutants has
caused loss of some organisms and accumulation of lead in others (Ward
and Brooks 1978).

Acid precipitation resulting from oxides of sulfur and nitrogen
can affect terrestrial ecosystems (Drablös and Tollan 1980), but the
extent of these effects is difficult to assess. The subject is
discussed more thoroughly in Chapter 8.

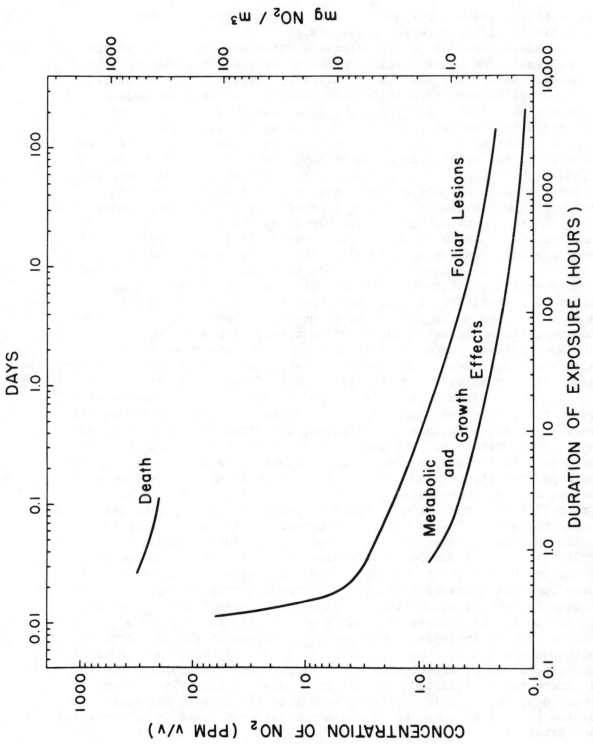

FIGURE 6.4   Thresholds for the death of plants, foliar lesions, and metabolic or growth effects as related to the nitrogen dioxide concentration and the duration of exposure.
SOURCE: MacLean (1975).

Aquatic Ecosystems

The effects of acid precipitation on aquatic ecosystems are well documented, and several symposia have recently been held on the problem (Shriner et al. 1980, Drablös and Tollan 1980). The effects are largely on organisms in poorly buffered lakes, where alkalinity is less than 100 microequivalents per liter. In such lakes, decreases in alkalinity and pH have been detected at several localities where the precipitation pH is 4.7 or less, both in Scandinavia (Henriksen 1980) and in North America. Chapter 8 gives a detailed account of the acid precipitation problem, and it will not be repeated here.

Trace metals and other trace substances are known to be toxic at concentrations found in soils and fresh waters within a few kilometers of smelters (Hutchinson and Whitby 1977). Toxic effects farther afield are more difficult to predict, because of inadequate background data. Apparently most of the trace organic pollutants transported via the atmosphere that have proved to be troublesome in ecosystems remote from sources are lipid-soluble. Examples include mercury in the methylated form, chlorinated hydrocarbons, and PCBs. In the upper Great Lakes, PCB concentrations in Coho and Chinook salmon, catfish, and eel are unacceptable for human consumption in some localities. Reproductive failure and deformities have been reported for herring gulls as a result (International Joint Commission 1977). The widespread dispersal of PCBs appears to have resulted largely from atmospheric inputs (International Joint Commission 1980, Murphy and Rzeszutko 1977). Concentrations of mercury in fish considered unacceptable for human consumption are often no more than 2 to 3 times natural concentrations, and the increase in release of mercury to the atmosphere caused by man is well over that (Lantzy and Mackenzie 1979). As mentioned in Chapter 1, cadmium and zinc are approaching toxic concentrations in Lake Michigan (Muhlbaier and Tisue 1980). Other lakes have not been studied, so it is not known how widespread this problem may be.

The toxicity of a substance to an organism may be displayed in different ways at different concentrations. For example, copper is known to affect behavior of freshwater fishes at concentrations as low as 4 micrograms per liter, but at this concentration there is no measurable chemical or physiological sign of stress. At 10 micrograms per liter, growth, reproduction, and mortality are affected. Blood chemistry and respiration are not affected until still higher concentrations. The effects of lead are known to be similar, with neurological damage occurring at lower concentrations than could be detected chemically or physiologically. Significant effects were noted at 8 micrograms per liter (Spry et al. 1981).

Interactions among Pollutants or between Pollutants and Other Stresses

The effects on the biosphere of many pollutants are influenced by other pollutants and elements normally present in the atmosphere, water, and soil. Sometimes the influences are indirect; for example, the solubility of many heavy metals is enhanced at low pH--i.e., high hydrogen ion concentration--and toxicity increases correspondingly. When direct interactions enhance the effects of one or both pollutants, the relationship is termed synergism; the result is more than additive. When interactions mitigate harmful effects, the relationship is termed antagonism.

Terrestrial Ecosystems

Menser and Heggestad (1966) were the first to demonstrate the synergistic effect of two gaseous pollutants upon plants. They reported injury to several tobacco cultivars when these cultivars were exposed simultaneously to concentrations of SO_2 and O_3 that individually would not injure the plants. Dochinger et al. (1970) and Houston (1974) reported a similar effect of SO_2 and O_3 in a tree species, Pinus strobus. Table 6.2 gives a summary of the synergistic effects of SO_2 with other gaseous pollutants on native North American trees. SO_2 and NO_2, and NO_2 and NO are known to have additive effects on the pea, Pisum sativum (Figure 6.5). Reinert et al. (1975) review pollutant interaction effects on terrestrial vegetation, including examples of synergisms and antagonisms. The authors point out that such pollutant interactions depend on the genetics and physiology of the target organism, the environmental conditions, and the concentration of pollutants--which may explain some of the contradictory findings reported in the literature. Holdgate (1979a) has made a more recent review; he believes different investigations have yielded different conclusions because of a lack of standardized methods and experimental conditions. Genetic, physiological, and environmental factors must all be standardized before exact comparisons are possible.

Interesting synergistic interactions have been described between ozone and both cadmium and nickel (Czuba and Ormood 1975), in their toxic effects on a number of horticultural crops. In contrast, the presence of copper in Sudbury area soils was reported to reduce the toxicity of fumigations by SO_2 for some grass species (Toivonen and Hofstra 1979). The mechanisms of these metal-gas interactions are not known. Effects through stomatal activity appear likely.

A number of authors have observed an increased susceptibility of forests to insect outbreaks after exposure to SO_2 and other atmospheric pollutants. This subject is reviewed by Renwick and Potter (1981) who attribute the infestations to an increased output of insect attracting terpenes.

Aquatic Ecosystems

High contents of hydrogen ion in freshwater bodies enhance the solubility of a number of elements normally present in soils and sediments. These include aluminum, manganese, zinc, and iron. Also

TABLE 6.2 Effects on Native Vegetation of SO_2 in Combination with Other Pollutants

Time	Concentrations of Gases in Mixture	Response of Vegetation	Plant Species	Reference
6 h/day for 28 days	0.14 ppm SO_2 0.05 ppm O_3 0.10 ppm NO_2	Significant reduction of growth (measured as height) compared with response to ozone and sulfur dioxide combined or ozone alone. Needles were significantly narrower than for any other exposure. This study is an example of growth reduction with slight foliar symptoms. Foliar response was most sensitive in early July.	Loblolly pine (*Pinus taeda*) (2 weeks old) Sycamore (*Datams occidentales*) (1 week old) seedlings	Kress and Skelly (1977)
6 h	0.025 ppm SO_2 0.05 ppm O_3	White pine needles 20 to 28 days old exposed 9 a.m. to 3 p.m.; response judged in terms of needle elongation (growth) and foliar lesion or tip necrosis. Author calls response synergistic. 0.025 ppm SO_2 or 0.05 ppm O_3 = threshold; 0.10 ppm O_3 = 20 percent necrosis.	Eastern white pine (*Pinus strobus*)	Houston (1974)
4-8 h/day, 5 days/week, 4-8 weeks	0.10 ppm SO_2 0.10 ppm O_3	16 percent needle necrosis (chlorotic, yellow spots, current year needles thin and twisted); shedding of older needles was far more extensive than in responses to single exposures. 0.1 ppm SO_2 injured 4 percent of needle area; 0.1 ppm O_3 injured 3 percent of area.	Eastern white pine (*Pinus strobus*)	Dochinger et al. (1970)

SOURCE: After Glass and Loucks (1980).

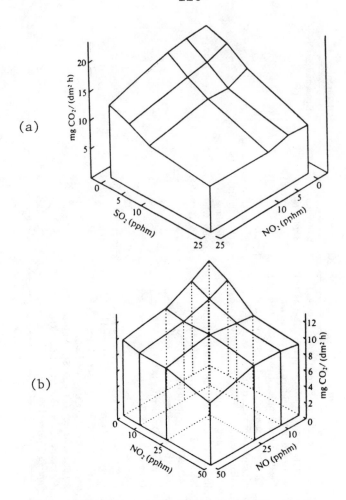

FIGURE 6.5   Interactive effects of pollutants. (a) Effects of SO_2 and NO_2 pollution on the rate of net photosynthesis in pea, *Pisum sativum* (from Bull and Mansfield 1974). (b) Effects of NO and NO_2 pollution on the rate of net photosynthesis in tomato (from Capron and Mansfield 1977). Note that in both cases combinations of pollutants have a greater effect than either gas does alone. SOURCE: Holdgate (1979b).

increased is the toxicity of metals, including copper, lead, zinc, and nickel (Spry et al. 1981). The presence of elevated calcium at low pH, however, reduces the otherwise high toxicity of aluminum--i.e., it acts as an ameliorative for both terrestrial higher plants and for aquatic freshwater plankton (Hutchinson and Collins 1978). This appears to be due to competitive inhibition of aluminum uptake by calcium. Higher calcium concentrations also appear to cause Crustacea to be less sensitive to hydrogen ions (Figure 6.6).

Among toxic metals a rather wide range of interactions occur. Nickel and copper in lake waters contaminated by the Sudbury smelters act synergistically in their effects on various green algae (Hutchinson 1973, Hutchinson and Stokes 1975). Stokes (1975) has shown that this effect relates to the initial separation of copper and nickel in the cell, with copper moving across the plasmalemma while nickel is bound to the cell wall. Eventually, sufficient damage occurs to the cell membrane that it becomes leaky and nickel enters also, this being the synergistic phase. Nickel-copper synergism has also been described for higher plants (Whitby and Hutchinson 1974). In studies of the floating aquatic plants Lemna valdiviana and Salvinia natans it was found that, in addition, cadmium and zinc also acted synergistically. Frond numbers increased with the tissues' concentrations of the two elements, and these were mutually enhanced by the presence of the second element (Hutchinson and Czyrska 1973). Both light intensity and the specific range of concentration influenced the outcome.

Water quality criteria based on laboratory tests of single pollutants appear to be inadequate for situations where several pollutants enter a water body. Wong et al. (1978) found that a mixture of heavy metals added to Great Lakes water or culture medium at the maximum permissible concentration levels for each was very toxic to freshwater algae. Diatoms were even more sensitive than either green or blue-green algae. Similar findings were reported by the MELIMEX study in Switzerland, where a mixture of metals was added to lake water at maximum permissible concentrations for each. Toxicity to phytoplankton (Gächter and Mares 1979) and heterotrophic microorganisms (Bossard and Gächter 1979) was reported for the mixture of metals, although singly the metals caused no observable damage. On the other hand, the metals caused populations of benthic organisms to increase, presumably because inhibition of the decomposer microbiota allowed more oxygen to accumulate at the mud-water interface (Lang and Lang-Dobler 1979).

Another example of synergistic effects of metals in animals is the strong teratogenicity of the combined injection of cadmium and lead compared to the effects of administration of either of the metals separately.

In contrast, metal interactions can protect against the normal toxic effects of a single metal. Such interactions have been described for mammals and fish (Cherian and Goyer 1978, Nordberg 1978, Parizek 1978, Levander 1977, Groth et al. 1976, Ganther et al. 1973); for invertebrates (Talbot and Magee 1978); and for unicellular green algae (Hutchinson 1973). The biochemical mechanism of the protective

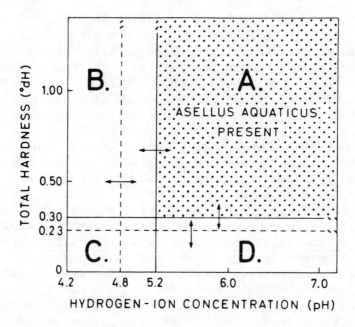

FIGURE 6.6 A model for tolerance limits of *Asellus aquaticus* toward low water hardness and low pH. The species is absent in lakes belonging to areas B, C, and D. Arrows indicate that limits may be moved through synergism with other factors. One degree dissolved hardness (1° dH) = 10 mg CaO/1. SOURCE: Ökland (1980).

effect is often the induction of metallothionein, which acts
nonspecifically to chelate metals (Kagi and Nordberg 1979). For
example, exposure of an organism to any one of cadmium, mercury,
silver, copper, or zinc causes increased tolerance to any of the other
metals upon subsequent exposure (Winge et al. 1975). When the
protective effect involves selenium (Ganther 1978), the ameliorative
action is thought to involve the diversion of metals from the
biochemical sites of toxic action (Bark et al. 1974, Komsta-Szumska et
al. 1976).

Previous exposure also affects the organisms' response to toxic
concentrations of metals. Some metals, such as silver, arsenic,
cadmium, mercury, lead, and tin are less toxic when sub-lethal doses
are given prior to a dose that normally produces mortality. Others,
such as barium, chromium, and iron, have toxicity enhanced by
pretreatment (Yoshikawa 1970). In cases where tolerance increases,
the underlying mechanism may involve metallothionein (Leber and Miya
1976, Probst et al. 1977), selenoproteins (Sandholm 1974, Prohaska and
Ganther 1977), glutathione (Congiu et al. 1979), or other mechanisms
(Tandon et al. 1980). Increased tolerance to metals may also be due
to physiological mechanisms, such as increased excretion of metals
induced by pre-exposures (Levander and Argrett 1969, McConnell and
Carpenter 1971).

A mixture of detergents and mercury or cadmium was found by
Calamari and Marchetti (1973) to have effects on rainbow trout
different from the effects of detergents or metals alone. Anionic
detergents plus metal had more than additive effects, while nonionic
detergents plus metal had less than additive effects. On the other
hand, phosphorus and nitrogen were found to increase the tolerance of
algae for heavy metals, as demonstrated in the tolerance of
Stigeoclonium tenue Kutz to heavy metals in South Wales (McLean 1974)
and the increased tolerance of microbiota and algae to arsenic in
waters enriched with nutrients (Brunskill et al. 1980).

As is well known from experience with human illnesses, stress
often renders organisms susceptible to diseases of various types. For
example, Hetrick et al. (1979) found that exposure of rainbow trout to
copper increased their susceptibility to infectious hematopoetic
necrosis and other viral diseases. The general relationship between
pollution stress and the incidence of infectious diseases in fish was
reviewed by Snieszko (1974).

In the marine environment, mercury contamination is accompanied by
enhanced concentrations of selenium in all investigated species of
mammals, birds, and fish--possibly due to a normal homeostatic
regulation. It seems likely that selenium exerts a protective action
against mercury toxicity in the marine environment, decreasing the
detrimental effects of mercury on reproduction behavior, growth, etc.,
and thus protecting the population and the ecosystem. In freshwater
ecosystems, high selenium retarded the uptake of mercury by freshwater
organisms (Rudd et al. 1980). On the other hand, Beijer and Jernelov
(1978) have shown that high selenium increases the retention of
mercury by aquatic organisms leading to bioaccumulation of mercury in
fish and a higher body burden in the individual. This might

counteract the positive effect of selenium in decreasing mercury uptake and toxicity.

In the environment, the presence of several pollutants and other kinds of stress is the norm. We must develop a fuller understanding of the mechanisms of interactions. Often, synergisms and antagonisms may occur simultaneously, so that no apparent effect may be detectable. For example, Eaton (1973) exposed Pimephales promelas to a mixture of copper, cadmium, and zinc for one year. First analyses showed the mixture to be no more toxic than zinc alone had been in earlier studies. When various effects were evaluated, however, it appeared that zinc toxicity was unchanged, while copper toxicity had increased and cadmium toxicity had decreased.

Resistance to Pollutants

There is considerable evidence that both individuals and populations can develop resistance to a wide variety of pollutants. At the individual level, increased resistance appears most often to be acquired through enzyme induction and the appearance of toxicant-binding proteins. The degree of resistance that can be acquired appears to vary both from one species to another and one toxicant to another, ranging from no increased tolerance to increases of several thousand times.

Resistance to several toxic metals by a wide variety of aquatic organisms appears to be the result of synthesis of metallothionein (G. Brown 1976), a protein that reportedly binds metals to sulfhydryl groups and prevents them from diffusing across cellular membranes into sensitive tissues or interacting with critical enzymes (Cherian and Goyer 1978, Brown and Parsons 1978). This protein develops in the organism in response to exposure to sublethal concentrations of several metals--Hg, Cd, Cu, Zn, Ag, and Sn--(Winge et al. 1975) after which resistance to a number of metals is increased. The amount of resistance that can be developed is high in most fishes but low in zooplankton (Brown and Parsons 1978). The protein has been shown to occur in most phyla of plants and animals (Brown 1980).

Similar multipollutant resistance appears to occur with pesticides and chlorinated hydrocarbons. Exposure to sublethal levels of one chlorinated hydrocarbon often increases tolerance to others. In the case of DDT, three mechanisms have been found to be involved: (1) a membrane barrier to DDT passage, (2) a structural difference in myelin, and (3) an efficient blood-brain barrier. Wells and Yarbrough (1972) and Moffett and Yarbrough (1972) discuss the details of these mechanisms.

A high intrinsic genetic variability in a population increases its chances of having some resistant survivors to a broad range of pollutant stresses. Species that occupy a wide variety of habitats under natural conditions tend to have high genetic variability. For example, the ubiquitous algal genera Chlorella and Scenedesmus will survive concentrations of heavy metals lethal to most other algal species (Stokes et al. 1973).

Structures or behaviors of some organisms protect them from atmospheric pollutants (Jacobson 1980). Waxy cuticles, surface hairs with hydrophobic properties, flat or narrow shapes, and vertical orientation tend to protect leaf surfaces from penetration of aqueous solutions of pollutants. Reproductive organs may be protected by flowers with inverted openings or flowers that close during cloudy periods. Other flowers open only when triggered by the weight of insects. Duration of pollination and number of pollinating flowers are also considerations. Salts on leaf surfaces or internal buffering reactions may also offer protection (Table 6.3).

Different genotypes in single plant or animal species are known to vary in their susceptibility to pollutants, as do different species within the same genus (Taylor 1978). Animal populations also evolve characteristics consistent with a polluted environment. For example, in industrial areas of Britain, elimination of pollutant-sensitive lichens from the stems of trees caused the disappearance of light-colored peppered moths, which had relied on the lichens for camouflage from insectivorous birds. A previously rare soot-colored "melanic" variant of the same species replaced the light form as the dominant genotype. The melanic form was also better adapted to eating pollutant-laden vegetation (Kettlewell 1955, 1956).

Some resistance, including that of the melanic peppered moths and tolerance of ryegrass to coal-smoke pollutants, is known to have developed within the past century or two. The gene pools of organisms are well prepared to adapt to changing environmental conditions, as long as the changes occur slowly enough for adaptation to occur. This is usually a function of the length of the generation time for a given species. Microbes can complete a life cycle in hours; larger mammals and tree species may take decades to centuries.

In areas of airborne fallout of contaminating metals and on old mine waste dumps, numerous examples of genotypes highly tolerant of normally toxic metal levels have been found. Such selection has been reported for a wide variety of higher plants, ferns, green or blue-green algae, bacteria, and fungi (e.g. Bradshaw 1952; Jowett 1958; Stokes et al. 1973; Tatsuyamo et al. 1975; Allen and Sheppard 1971; Cox and Hutchinson 1979, 1980) and for a number of metals and combinations of metals.

Past Efforts to Predict Widespread Toxic Effects

Both scientific induction and unplanned-for catastrophes have led to an understanding of biological effects of pollutants. Perhaps most illustrative is the example of DDT. Before its extensive broadcast about the environment as an insecticide following World War II, its potential effects upon nontarget organisms were recognized by several scientists, including Cottam and Higgins (1946). They pointed out that agricultural applications would affect wildlife and game and were especially concerned about its entry to streams, lakes, and coastal bays, where the crabs and fish would be sensitive to this toxic halogenated hydrocarbon.

TABLE 6.3   Plant Processes and Characteristics That May Increase Tolerance to Acid Precipitation

Exclusion
  Leaf orientation and morphology
  Chemical composition of cuticle
  Flower orientation
  Protection of sexual organs
  Pollination mechanism
Neutralization
  Salts on leaf surfaces
  Buffering capacity of leaves
Metabolic Feedback Reactions
  Enzymatic reactions that consume hydrogen ions or yield alkaline products

SOURCE: Jacobson (1980).

Subsequent events have confirmed that higher trophic levels are jeopardized by exposure to DDT. One of the more publicized effects upon the marine ecosystem involved the reproductive failures of the brown pelican population on Anacapa Island, off the California coast, from 1969 to 1972. The accumulation of DDT and its degradation products, primarily DDE, in marine organisms that were the food for the pelicans initiated the problem. The source of the pesticide was allegedly wastes from a manufacturing plant in Los Angeles, which were introduced to the oceans in sewer outfalls. The pelicans produced thin, friable egg shells, which broke easily (Risebrough 1972). DDT usage has been severely restricted since the early 1970s by the United States and many nations of the Northern Hemisphere. The brown pelican population on Anacapa Island has shown breeding recovery since 1972 when the DDT dumping was stopped. Despite DDE residues in the sediments, the success rate for young per nest had risen from 0.04 in 1972 to 0.88 in 1975 (Schreiber 1980). This recovery is true in other locations also.

With nuclear weapons development in the early 1950s, scientists from several countries became concerned about the entry of artificially produced radionuclides from power plants to the atmosphere and to the oceans. The ingestion of food containing radionuclides from the sea and the exposure of individuals to environmental radioactivity were minimized through limitations of the amounts of wastes emitted from nuclear plants to their surroundings. The greatest amounts of artificial radionuclides introduced to the environment come from the nuclear processing plant at Windscale, England. British environmental scientists have developed protocols to regulate the discharges on the basis of the "critical pathways technique," through which nuclides posing dangers to human health are identified and the critical population for exposure is protected. As a consequence of the resultant regulatory measures, there appears to be no danger to the population liable to exposure at the present time.

SUMMARY

Accumulation of pollutants in ecosystems is affected by a number of physical factors at the surface of the ecosystem, such as the amount of surface area presented by a forest canopy or turbulence near the air-water interface. The form of the pollutant--gas, aerosol, or large particulates--is also important. Ice and snow cover and thermocline formation may protect freshwater ecosystems from dissolved pollutants at certain times of the year, but can cause higher than average inputs at others. The thermocline regulates the passage of pollutants to and from the deep ocean.

Soils and aquatic sediments tend to accumulate many pollutants, thereby lowering toxicity to pelagic or above-ground forms but raising toxicity for benthic or soil organisms. The binding in soils and sediments for some elements depends on oxidation-reduction potentials, which are lowered when anoxic conditions develop. Many organic constituents are transformed by microbes in soils or sediments into other organic compounds, which might be either toxic or harmless, or into inorganic products, CO_2 and H_2O.

Substances that are not degraded in terrestrial ecosystems will eventually be transferred to aquatic ecosystems via water flow or soil erosion; hence, higher concentrations of toxicants are likely to occur in aquatic ecosystems. Aquatic sediments often provide a reservoir of datable toxic deposits and fossil organisms, from which a history of pollution might be reconstructed.

At the species or individual level, pollutants may be bioconcentrated, bioexcluded, or biomagnified by passage up the food chain.

Gaseous pollutants have their greatest effect on terrestrial ecosystems, because the air-water interface offers to aquatic organisms some degree of protection from episodic events.

When many pollutants are introduced to an ecosystem, effects are in many cases synergistic--i.e., effects are greater than additive. In some other cases less than additive effects have been noted.

Natural selection has caused some species to develop resistance to pollutants. Such genetic adaptation requires many life cycles, and no examples are known for species with life cycles over one year long. Individual resistance may, however, develop in individuals exposed to sublethal concentrations of pollutants.

7. STUDYING THE EFFECTS OF ATMOSPHERIC DEPOSITION ON ECOSYSTEMS

The widespread dispersal of trace pollutants--many of them toxic, carcinogenic, or mutagenic--has been discovered only recently, through tremendous advances in the technology of chemical analysis in the past two decades. These technical developments have increased the sensitivity of detection of most chemicals by orders of magnitude. The methods and programs available for assessing the environmental and biological effects of these trace atmospheric pollutants, however, have not reached a comparable degree of sensitivity or reliability. Substances are usually tested for their effects over short periods--rarely as long as one complete life cycle--and at relatively high dose rates, either singly or in very simple combinations with other pollutants. Often such tests are inadequate to assess long term, chronic problems. For example, it took thirty years of careful statistical studies of human populations, combined with experimental studies of animals, to document the linkage between smoking and lung cancer (U.S. DHEW 1979).

There are many examples of unexplained mortalities that could well be caused at least partially by multiple pollutant stresses. Among them are the decline of leopard frog populations in Minnesota (McKinnell et al. 1979), the high cancer mortality in workers in the petrochemical industry (Thomas et al. 1980), and the high incidence of lip cancer in sucker fish in Lake Ontario. Epizootics of benign skin tumors (papillomas) were found in white sucker, Catastomas commersoni populations throughout the Great Lakes. However, elevated tumor frequencies (50.8%) were found in populations clustered around an industrial complex on Lake Ontario concerned with petrochemical refining. In this region there was an anatomical shift in the tumor location to the lips, an anatomical site which has near constant contact with bottom sediments (Sonstegard and Leatherland 1980). Disentangling the chain of cause and effect responsible for such mortalities is an extremely difficult but extremely important task. Mortalities in animal or plant populations may serve--as did the coal miner's canary--to provide early warning of human hazard.

Linkages among pollutants are also of critical importance in determining how whole ecosystems are affected. Oxides of sulfur and nitrogen are strongly linked in coal combustion, and both are linked to trace metals and trace organic molecules (e.g., polycyclic aromatic hydrocarbons) produced by combustion. In metal smelters there is less linkage between SO_x and NO_x and a much stronger link between SO_x and metal emissions. In vehicle exhaust, NO_x predominates strongly over SO_x and is strongly linked to trace organic materials. The total effect of the linked emissions on an ecosystem depends on the combination of pathways and interactions of the individual emissions.

Interactions along ecosystem pathways are especially diverse. For example, sulfate and nitrate deposited from the atmosphere to terrestrial ecosystems may travel together as rainwater percolates through soils, while hydrophobic and lipid-soluble organic molecules will follow other pathways. Because of greater biological demand and generally smaller supply, a greater proportion of nitrate than sulfate may pass into storage in biotic compartments of the ecosystem, leaving relatively more of the sulfate to pass through to streams and lakes. In aquatic or wetland habitats some sulfate may be reduced to sulfide in anaerobic environments and end up deposited as ferrous or other metallic sulfides in a sedimentary sink. Both nitrate and sulfate may be taken up by organisms and reach the sedimentary sink by a detrital pathway. Alternatively, some sulfate in anaerobic habitats may be reduced to volatile sulfides, and some nitrate to gaseous ammonia, which return to the atmosphere, where the sulfides may again be oxidized and interact with ammonia to form ammonium sulfate aerosols.

Metals such as lead, which often accompany sulfate in particulate pollution, will follow yet another path, becoming strongly adsorbed to ion-exchange sites on soil clays or humic acids. Mercury, another metallic component of particulates from coal combustion, may become methylated by sedimentary microorganisms and thus be returned to the atmosphere as volatile dimethyl mercury; it can also volatilize as metallic mercury (Wood 1974). Still other metals may be mobilized from soils by acidification due to atmospheric depositions; among these are nutrient elements such as calcium and potassium and toxic elements such as aluminum and zinc. Thus, not only are individual organisms affected by combinations of pollutants, but the ecosystem as a whole is affected by the combination of pollutants interacting with different parts of the system in various ways.

Current knowledge of atmosphere/biosphere interactions suggests a number of broad-scale hypotheses concerning future ecological damage that could result from air pollutants. Of greatest global, long-term significance is the hypothesis that atmospheric pollutants such as carbon dioxide and oxides of nitrogen released in the course of energy generation may warm the planetary atmosphere (NRC 1977a, Kellog and Schware 1981). The consequences of such a "greenhouse effect" for the earth's terrestrial and aquatic biota are at present unpredictable, but major shifts in the global patterns of plant productivity and human habitation seem very likely (Kellogg 1978).

A second major hypothesis is that anthropogenic release of oxides of sulfur and nitrogen may alter appreciably the cycles of sulfur and nitrogen, again with consequences for the biota that are difficult to predict. On the one hand, both elements are plant nutrients that are deficient in certain soils; on the other hand, both contribute to "acid rain" (more correctly acid deposition), which is already a serious regional problem in North America and Europe (Dochinger and Seliga 1976, Hutchinson and Havas 1980, Drablös and Tollan 1980). In addition to directly affecting the biota, acid precipitation can alter substantially the cycling of both nutrients and toxicants in acid-sensitive ecosystems, particularly those on crystalline Precambrian rocks. A most recent and unexpected manifestation of acid-rain effects is the acidification of ground waters in Scandinavia (Hultberg and Wenblad 1980), discussed in more detail in Chapter 8.

A third, less well supported hypothesis is that energy-related air pollution is becoming a serious cause of biotic impoverishment. Increased acidification of fresh water streams and lakes by acid precipitation has destroyed sensitive populations of fishes and other aquatic organisms. Fallout of toxic heavy metals from the atmosphere and gaseous oxides of sulfur and nitrogen have been shown to have similar effects. Even if losses are only of local plant or animal varieties, a serious loss of genetic diversity may result. As pollution continues and increases, the effects will become more widespread and more severe.

One of the most difficult problems to assess is the possibility that chronic low-level pollution of the atmosphere by teratogens, carcinogens, and mutagens is affecting the health, reproductive capacity, survival, and disease resistance of the biota, and thus indirectly the very structure and function of the biosphere that is our life-support system. By contaminating the environment and by causing the extinction of sensitive genotypes, air pollutants, acting singly or in combination, may change the biosphere in ways that are essentially irreversible. The possibility of major alterations in planetary ecology and biogeochemistry raises important questions concerning the kinds of programs that are necessary to predict, investigate, and cope with the problems.

PREDICTING ANTHROPOGENIC EFFECTS ON ECOSYSTEMS

Separating human-induced ecological changes from natural ones and measuring them is not a simple matter, particularly when the changes occur slowly. In describing such changes and in establishing causality, there is often an implicit assumption that in the absence of human intervention ecosystems are in some sort of distinguishable and predictable successional state. Ecological theory assumes that following small perturbations, an ecosystem will return to the original successional pattern. This is a useful theoretical concept but a difficult one to apply to particular ecosystems. Clearly distinguishable ecosystem successions are hard to predict even when human interference is absent. The maximum perturbation that an ecosystem can tolerate without irreversible change is not generally known. Moreover, there may be more than one possible successional pattern, even under natural conditions. On a long time scale,

climatic oscillations will influence the nature of the successional
state. Superimposed on this oscillating system is the natural
tendency of some species to become extinct while new ones develop.

Human-induced changes must be considered against the background
of a normal range of natural trends and fluctuations. The system with
which we must deal is shown, greatly simplified, in Figure 7.1. Let
us review, then, the types of information necessary for predicting the
biological effects of atmospheric deposition.

- For predictive purposes, one would like to know which source
  characteristics and emission processes govern the various
  substances emitted, their forms and properties, and their
  rates of emission. Anthropogenic sources and their emission
  rates are readily measurable, but natural fluxes are
  difficult to estimate.
- Although the relationship between transport distance and
  residence time for particles of different size and type is
  reasonably well understood (Figure 5.1), much remains to be
  learned about pollutant transformations in the atmosphere.
- The interfaces between atmosphere, lithosphere, hydrosphere,
  and biosphere (Figure 7.1a) are of critical importance in
  elucidating the ecological effects of pollutants. Transfers
  across them are controlled by the physical and chemical
  properties of the pollutants (or their alteration products),
  by the characteristics of the receptors, and by the nature
  of the transfer processes. Some aspects of transfer, such
  as wet deposition from the atmosphere, are relatively well
  understood; others, such as dry deposition, are not.

Some important pollutant transfers and their effects are
relatively simple to follow. For instance, SO_2 transfer to plant
leaves (A → O in Fig. 7.1a) causes a direct visible injury. However,
SO_2 may also exhibit a variety of indirect effects. By transfer to
soils and oxidation to H_2SO_4 it may affect cycles of nutrients and
toxicants and thus plant growth (A → S → O in Fig. 7.1a). By
transfer direct to waters, as H_2SO_4 in snowmelt percolating over
frozen ground, it may result in fish kills (A → W → O in Fig.
7.1a). Or by transfer to soils and then to waters as H_2SO_4, it
may inhibit fish reproduction (A → S → W → O in Fig. 7.1a). It may
even react through organisms back to the environment, as when SO_2
fumigation kills the vegetation around a metal smelter, leading to
severe soil erosion (A → O → S in Fig. 7.1a).

- For each of the cases mentioned above, attention will have
  to be paid to all of the aspects of pollutant transfer shown
  in Fig. 7.1b. The transfer of a pollutant within and
  between the diverse terrestrial and aquatic compartments of
  ecosystems is extremely complex, depending upon the
  characteristics of both the pollutant and the numerous
  receptor sites and transfer processes within an ecosystem.
  Interactive effects upon the various species that dominate
  different trophic levels are especially difficult to
  predict, because each receptor species will respond
  differently (in greater or lesser degree) to a given
  pollutant and also to its effect upon other species.

123

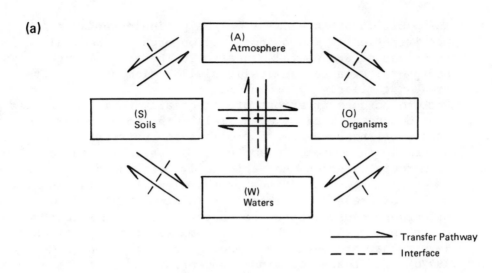

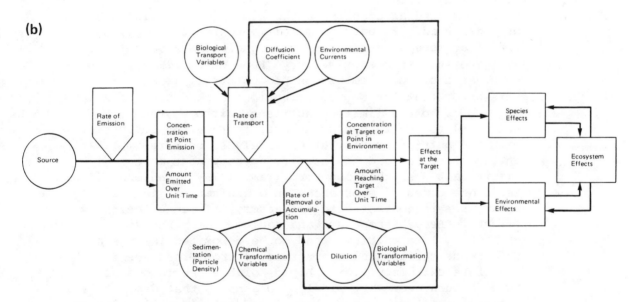

FIGURE 7.1 Models of interactions in the atmosphere-biosphere system. (a) Global interactions between major biogeochemical reservoirs. (b) Generalized pathways for atmospheric pollutant effects. SOURCE: After Holdgate (1979b).

- The position of an element on the periodic table or its participation in known inorganic or organic chemical reactions offers useful clues to its behavior in the biosphere, because chemically similar elements are often treated similarly by organisms.

- Background concentrations of potentially toxic substances must be known, because even small anthropogenic releases of such substances may cause problems if the substance occurs naturally in near-toxic amounts. For example, the doubling of mercury emissions has apparently been enough to cause undesirable concentrations of the element in fish in a number of areas. The increases in atmospheric nitrous oxides and hydrocarbons from anthropogenic emissions can cause an increase in the photochemical formation of ozone to levels that are severely damaging to vegetation (NRC 1977b). Increased aluminum concentrations in surface water caused by acid deposition have also proved detrimental. Substances not normally found in near-toxic concentrations but released in large quantities by anthropogenic sources may also be problems. Examples are lead and oxides of sulfur and nitrogen.

- Synergisms and antagonisms among atmospheric pollutants must be considered. Effects of metallic toxicants (e.g., Hg, Al) are often more severe under acid conditions. Selenium may mitigate the effects of mercury (Rudd et al. 1980); both are components of coal combustion products. It is also noteworthy that the NO_x contribution to acid precipitation may supply both a limiting nutrient, nitrogen, and a potential toxicant, hydrogen ions, to receptor ecosystems, with the nutrient contribution likely to be more important in the short run and the acidifying component more significant over the long term (Tamm 1976, Abrahamsen 1980).

- Residence times of substances in organisms, or in the substrates in contact with organisms, influence their toxic effects. Lead, strontium, and radium, which behave chemically like calcium, tend to be deposited in bone or calcareous shells, and their removal from organisms is therefore extremely slow. Lipid-soluble substances also tend to turn over very slowly. Organisms that deposit large amounts of fat will accumulate lipid-soluble pollutants most strongly; birds that accumulate great amounts of fats prior to migration often contain large quantities of chlorinated hydrocarbons and PCBs which can be released to the blood stream as the fats are used up. Hutchinson et al. (1979b) have shown that the degree of toxicity of a wide range of hydrocarbons, including chlorinated hydrocarbons, to freshwater green algae is predictable with some accuracy, based on a number of key physico-chemical parameters of the hydrocarbons. Aqueous solubility is one such parameter and the octanol-water partition coefficient is another, both simulate aqueous-lipid partitioning. Infiltration into the cellular membranes is also predictable, as detected by K and Mn leakage, based on these same molecular characteristics. Kenaga (1980) has shown that bioacccumulation in cattle and swine is highly correlated with bioconcentration factors in fish. He notes that bioconcentration factors in beef fat are negatively correlated with aqueous solubility and are positively correlated with 1-octanol-water partition coefficients. Radioisotopes such as radiocarbon and tritium, which are components of compounds subject to much greater rates of turnover in organisms, appear in general to be less harmful than substances that turn over slowly, at least in the short term--excluding genetic effects.

- Knowledge of food chains is important in tracing biomagnification. Top predators--i.e., organisms at the top of the trophic pyramid--will often concentrate most strongly substances that are affected by biomagnification. Predatory fish, for example, accumulate both chlorinated hydrocarbons and mercury.

- Information on distribution and abundance of sensitive receptors and distribution and size of pollutant sources is important for determining the effect of a particular pollutant upon a particular ecosystem. A maximum effect could be expected in a situation in which the receptor species is highly sensitive to a particular pollutant, the species is rare but ecologically important, and either the sources and receptors are located together or the pollutant has a long atmospheric residence time.

Although we cannot provide foolproof prescriptions for the detection and evaluation of all pollutants, several general features characterize large numbers of undesirable pollutants, making it possible to categorize them to some degree. For instance, pollutants may be linked together at their source or share common transport or uptake mechanisms. They may also have the same targets in the biosphere, and exert similar effects upon them. A "comparative anatomy" constructed around similarities and differences among pollutants would reduce the number of surprise pollutants in the future, although there are always likely to be some pollutants with characteristics so unique as to defy prediction, or new categories of anthropogenic pollutants with properties vastly different from current ones.

Table 7.1 is a detailed summary of the information required for an ecological assessment of a pollutant. This table reveals the characteristics that can make an atmospheric pollutant dangerous. First there must be significant emissions (A.1) of a substance with considerable effect upon organisms (E.1, E.3), ecosystems (F.), the environment (G.), or man-made materials (H.). Long residence time in the atmosphere (B.4) favors its spread but dilutes the amount deposited in any area. Resistance to transformation (B.2) and detoxification by organisms (E.2) lengthen its persistence in the atmosphere and increase its resistance to degradation along ecosystem pathways (D.4). However, some pollutants (e.g., inorganic Hg) are

TABLE 7.1 Major Factors to be Determined in an Assessment of the
Ecological Effects of a Pollutant

A. Emissions

    1. Significant sources (coal, oil, natural gas, nonfossil fuels,
        natural processes)
    2. Nature of source (stationary, mobile)
    3. Nature of primary emissions (gases, particulates in different
        size fractions)

B. Atmospheric transport

    1. Significant modes (gases, particulates in different size
        fractions)
    2. Physical and chemical transformation (type, degree)
    3. Nature of transport (local, regional, global)
    4. Residence times (up to days, weeks, months, longer)

C. Deposition

    1. Dry (gaseous absorption, particle collision, settling)
    2. Wet (rainout, washout, as rain or snow)

D. Ecosystem pathways

    1. Readily transported (by water, or by volatilization back to
        atmosphere)
    2. Likely to accumulate in sinks (e.g., soils, sediments)
    3. Likely to accumulate in organisms (bioconcentration,
        biomagnification)
    4. Readily degradable (abiotically, biotically)

E. Biological consequences for microbes, plants, animals; in
    terrestrial, aquatic, wetland habitats

    1. Toxicity (low to high, acute to chronic, affecting few to
        many organisms)
    2. Detoxification and/or repair capability (low to high)
    3. Type of effect (on health, development, growth, reproduction,
        phenology, behavior, heredity)

F. Ecological effects

    1. Energy flow (production, decomposition, storage)
    2. Biogeochemical cycling (limiting nutrients, toxicants)
    3. Structure (vegetation patterns, trophic levels, species
        composition, stability)

G. <u>Effects on properties of the physico-chemical environment</u>

    1. Atmosphere (radiation balance, water balance, gaseous and particulate composition, visibility)

    2. Hydrosphere (thermal properties, water balance, chemical properties)

    3. Lithosphere (thermal properties, water balance, chemical properties)

H. <u>Effects on man-made materials</u>

    1. Economic (metals, other structural materials, fabrics, other organic materials)

    2. Aesthetic (works of art, architecture)

128

readily altered to products (e.g., methylmercury) that are even more toxic. And in the case of SO_2 and its alteration product H_2SO_4, the effects may be quite different, with the former affecting chiefly terrestrial ecosystems and the latter having a profound effect on sensitive streams and lakes (Glass and Loucks 1980). Solubility in water (D.1) favors mobility in the ecosystem and the avoidance of sinks in soils and sediments (D.2). Lipid solubility, on the other hand, favors entry into organisms by transfer across cell membranes (D.5) and both bioconcentration and biomagnification along food chains (D.3).

HIGHLY SENSITIVE ORGANISMS AS EARLY INDICATORS

Sensitive early indicators of pollutant stress may be useful in developing monitoring schemes to detect environmental stress in its initial stages. In most ecosystems, some species or groups prove much more sensitive than others. For instance, the bryophytes--i.e., mosses and liverworts--and lichens lack the waxy cuticle of the flowering plants, and thus their epidermal cell walls are directly exposed to the air. These cell walls have a highly charged cation exchange surface, and thus many deposited elements are retained by exchange.

Such organisms are usually the most sensitive indicators of pollutant stress. Whole assemblages of lichens and bryophytes have become rare in urban areas (Holdgate 1979a, Johnsen 1980). These organisms are so sensitive that their assemblages have been proposed as a general index of atmospheric pollution (Isecutant and Margot, discussed by Holdgate 1979a). Such organisms are also likely to accumulate trace substances. For example, trace metals have been shown to be concentrated from 1,000 to 50,000 times by bryophytes (Dvorak et al. 1978). Ruhling and Tyler (1968 and 1970) have used the feather moss Hylocomium splendens for assessing lead loadings near highways in Sweden, and for examining north-south latitudinal gradients. Cesium residence times in mats of Cladonia lichens were found to range from 4 to 25 years (Lidén and Gustafsson 1967).

Some aquatic plankton groups, such as the Chrysophyta or Crustacea, respond to minuscule quantities of pollutants (de Noyelles et al. 1980, Marshall and Mellinger 1980). Common responses include depression of photosynthesis and changes in species composition (de Noyelles et al. 1980). Changes such as the above may in turn affect other members of the ecosystem that depend on the displaced species for food or protection.

Among animals, Gammarus lacustris, a large benthic crustacean, and numerous molluscan species have proved more sensitive to acid precipitation than any other easily recognizable species (K.A. Ökland 1980). A North American animal with similar reactions, Mysis relicta, may prove even more useful. Like Gammarus, Mysis disappears when waters reach pH values of 5.8 to 6.0 or less (Schindler 1980a). The species is a glacial relict, and thus its distribution under earlier, unpolluted conditions is known quite exactly. Its absence in lakes that are otherwise suitable may be an indicator of acidification. The mud minnow, Pimephales promelas, is equally sensitive to acidification (Schindler 1980a, McCormick et al. 1980).

In general, embryonic development is the most sensitive stage in the life cycle of an organism (McKim 1977), and the incidence of deformity or biochemical abnormality at that stage may serve as an early warning for more far-reaching effects occurring at later stages of the life cycle. While teratological and immunological studies are now commonly done with mammals, they have seldom been done on animal or plant populations in nature. The few studies that have been done on natural populations have proved to be very sensitive indicators of pollutant stress (e.g., Kennedy 1981, Daye and Garside 1980). They may complement epidemiological studies of long-lived organisms to enhance our knowledge of the latent effects of long-term exposure to low concentrations of pollutants.

Relatively tolerant accumulator organisms may also be useful in monitoring the dispersion and persistence of dangerous substances. For example, mosses and lichens readily accumulate radioactive fallout (Gorham 1958a, 1959) and heavy metals (Ruhling and Tyler 1971), and a global "mussel watch" is now monitoring the dispersal of dangerous contaminants (such as organochlorides, trace metals, petroleum hydrocarbons, and transuranic elements) in coastal marine waters (NRC 1980b).

Prediction of sensitivity in organisms is sometimes difficult because a given organism can react in very different ways to different pollutants. For instance, lichens are exceptionally sensitive to SO_2 pollution but are unusually tolerant of radiation stress (Hawksworth et al. 1973). However, some reasonably useful generalizations can be suggested. In general, top carnivores are likely to be sensitive to persistent, fat-soluble pollutants capable of biomagnification (Carson 1962). Species with small chromosome volumes are insensitive to irradiation by radioisotopes (Woodwell and Sparrow 1963, Woodwell 1967). Opportunistic species that are colonists during early ecosystem succession also tend to be more resistant to pollutant stress than the usually longer-lived and more specialized organisms that inhabit relatively stable ecosystems. Presumably the opportunists are better able to tolerate diverse environmental conditions (Woodwell 1970) and also--because of high reproductive rates--to recover more readily from pollution damage.

COMMON RESPONSES OF ECOSYSTEMS TO STRESS

Ecosystems respond differently to different pollutants. For example, acid deposition affects chiefly aquatic ecosystems in watersheds with thin, coarse soils over crystalline igneous rocks. In this case, biotic effects are largely due to sensitivity of an abiotic compartment of the ecosystem. In contrast and as mentioned before, SO_2, the gaseous precursor of acid rain, affects directly and primarily the organisms of terrestrial ecosystems--no matter what the geological substrate--because in water it is rapidly converted to sulfate. The NO_x that contribute to acid deposition, however, may be expected to have their greatest direct effect upon ecosystems whose vegetation is limited by nitrogen. In a more general way, ecosystems with a complex structure based on long-lived species are likely to be especially sensitive to pollutant stress (Woodwell 1970).

Woodwell (1970) pointed out the similarities in the responses of deciduous woodlands to two apparently quite different stresses, gamma radiation and SO_2. He noted that in both cases large target trees close to a point source of either the radiation or the pollutant were badly damaged and that a zone of damage occurred. The most tolerant plants were sedges and grasses, both of which have their growing point--apical meristem--at or below the soil surface.

This pattern appears to be fairly general. Species diversity decreases as a result of such stresses as SO_2, radiation, fire, oil spills on land, extremes of cold (arctic and alpine), and high wind exposure or drought. The plants that seem best adapted to these stresses include a number of perrenial grasses and sedges. Annuals, which must complete a reproductive cycle each growing season, are very rare; large above-ground targets, such as trees and shrubs, are especially sensitive. Those species which have underground perennating organs such as bulbs and stolons or which are able to produce suckers from adventitious buds are likely to survive best. Vegetative reproduction is generally more successful under such stresses than sexual reproduction.

Bryophytes and lichens are especially tolerant of a number of these stresses--e.g., cold, wind, and radiation--but, as we have seen, where the stress involves entry of a pollutant into cell walls, the absence of a cuticle is a crucial disadvantage. Lichens and mosses generally are susceptible to atmospheric pollutants, such as SO_2, fluorine, and trace metals, which are absorbed on the exposed surfaces. They also may be susceptible to atmospheric hydrocarbons, as they are to direct application of petroleum. For example, of 31 species present initially at a tundra site in the Canadian arctic, 21 were bryophytes or lichens. But, one year after sprayed oil spills, all 21 of these species were dead, while 7 of the 10 higher plant species had survived. After 8 years, 9 of these 10 were present, but only 1 bryophyte had reappeared (Hutchinson 1981).

It should be noted that these generalizations about stress-susceptible or resistant vegetation do not apply to the successional aftermath. When the effects of the pollution episode have diminished, pioneer species may be able to invade the previously stressed area de novo.

Susceptibility of ecosystems to atmospheric pollutants seems to be reduced as annual precipitation is reduced, perhaps because toxicants can cause the greatest damage to organisms and populations that grow and metabolize rapidly, under conditions with optimum water and temperature. Emissions of SO_2 from smelters in desert regions such as Arizona have much less effect on vegetation than equal SO_2 emissions in temperate regions (Wood and Nash 1976). It therefore seems likely that rapidly growing tropical forests could sustain more severe damage from a given stress than temperate forests, and that slow-growing arid or arctic ecosystems would be less sensitive. There is no dormant period in the tropics, while in both arid and arctic regions very extensive dormant periods occur, during which above-ground photosynthetic material is greatly reduced.

Annual plants may be at special risk because they must successfully go through all the stages of sexual reproduction each year, and irregular episodes of exposure to harmful atmospheric pollutants could disrupt their reproductive cycle. One might predict that annuals in desert areas, where they are often abundant, could be locally eliminated while the perennials survive. Studies of the effect of smelters in plant populations in the Southwestern United States have confirmed this (Wood and Nash 1976).

A newly emerging problem is the possibility that expanding use of diesel fuel in the U.S. auto fleet may increase emissions of suspended soot particles by more than an order of magnitude (Barth and Blacker 1978, Springer 1978). This problem is currently being investigated by the Diesel Impact Study Committee of the National Academy of Engineering (NRC, in press). Diesel particles are mostly in the submicron size range, and thus are very effective in reducing visibility. Of greater ecological importance, they carry a variety of adsorbed toxic, carcinogenic, and mutagenic compounds such as benzo(a)pyrene, and they are small enough to penetrate deep into animal lungs. The ecosystems most likely to be affected are those close to heavily traveled highways. Among primary producers, epiphytic lichens could well be sensitive, for reasons mentioned above. The micro- and meso-fauna that are exposed to concentrated through-fall from the canopy and that process leaf litter along food chains in the soil humus layer might also be sensitive. If biomagnification of lipid-soluble organic pollutants takes place, then carnivores at the top of the food chain ought to be regarded as potentially sensitive. In aquatic ecosystems adjacent to busy highways, the neuston (water surface) dwellers may be vulnerable, because many organic molecules concentrate to an exceptional degree in the surface film at the air-water interface. In the water beneath, filter-feeding zooplankton may well collect soot particles more or less indiscriminately along with their food, as may detritivores in the benthos.

Cycle linkages may be of considerable importance in determining the total effect of air pollution upon a given organism and the ecosystem in which it occurs. For example, acid rain has both direct and indirect effects on fish; directly by toxicity of the hydrogen ion and indirectly by the mobilization of aluminum from soils. There is also a possibility, as suggested elsewhere in this report, that hydrogen ions may mobilize sufficient mercury from soils or sediments to become lethal as larger, older fish accumulate the element to high levels.

While this report has emphasized the potential significance of atmospheric deposition in terrestrial and freshwater ecosystems, it is important to note that similar concepts apply to estuarine and coastal marine environments. Considerable evidence is available documenting an enhancement in concentrations of heavy metals and organic compounds in estuarine and coastal environments by atmospheric transport from anthropogenic sources (Farmer et al. 1980, Bertine and Goldberg 1977, Seki and Paus 1979, Crecelius et al. 1975, Rodhe et al. 1980).

Understanding of the fate of anthropogenic pollutants delivered to the oceans by atmospheric transport is inadequate. Results from existing studies concerning the fate and ecological effects of anthropogenic pollutants show that the toxicity of trace metals to marine organisms is related to both concentration and chemical speciation (Steemann and Wium-Andersen 1970, Harriss et al. 1970, Allen et al. 1980). Elevated concentrations of certain trace metals in seawater--some at concentration levels that can be found to occur in some marine ecosystems--have been shown to be toxic to estuarine and marine organisms in both laboratory culture and enclosed water-column experiments (Eisler 1973, Eisler and Wapner 1975, Grice and Menzel 1978). A high priority should be given to research on analytical techniques, bioassay methodologies, and monitoring strategies for assessing the effects of both inorganic and organic pollutants in estuarine and coastal environments. Two recent workshop reports identify specific research needs (Goldberg 1979a,b).

NATURAL RECORDS OF POLLUTION

Records of the deposition of pollutants from the atmosphere in a given area may be found in the sedimentary record of lakes, reservoirs, marine deposits, and glaciers, and in annular deposits in trees and corals, if certain criteria are met. The processes that remove the pollutant from the atmosphere and the atmospheric residence time must have remained the same over the time interval of interest. The age assignments of the sedimentary strata must be accurate. And there must be no movement of the pollutant within the column following deposition.

Anoxic sediments are especially attractive for assessing pollutant concentrations, especially if the sediments lie under anoxic waters. In such cases there is practically no disturbance by organisms to smear the record through movements of sediments. In many areas the solid nature of glacial sediments helps maintain the various strata as closed systems with respect to the migration of pollutants.

The sedimentary record integrates the outputs of a variety of sources. For example, heavy metals may be introduced to the air by fossil fuel combustion, cement production, smelting, and many other activities. Although the relative contributions clearly are of great importance, knowledge of the overall atmospheric level and rate of increase is invaluable for regulatory activity.

Sediments from southern Lake Michigan have been studied to determine the atmospheric burdens, sources, and rates of increase of several trace metals emanating from adjacent highly populated, highly industrialized and highly agricultural areas (Goldberg et al. 1981). Sources were known to be primarily combustion processes involving coal, oil, and wood. Time horizons were determined for the Lake Michigan sedimentary column by the use of Pb-210 and Pu-239 + 240 chronologies. Primary energy sources in each time horizon were

identified by analyzing the characteristics of charcoals deposited in
the strata (Griffin and Goldberg 1979). This charcoal record shows
that in the 19th century (1830 to 1900), the primary combustion
processes involved natural and man-initiated burning of wood. This
changed in the beginning of the 20th century, when coal became the
important energy source. On the basis of the characteristics of
associated charcoal deposits, oil is thought to have been responsible
for pollutants in strata deposited after 1928. The charcoals
indicated that in the period 1953 to 1978, about 76 percent of
deposited charcoal had an origin in coal burning, 14 percent in oil
burning, and 10 percent in wood burning. There is a steady increase
of deposited carbon to about 1960 and then a distinct decrease to 1976
(Figure 7.2).

Nine of the twelve metals assayed (Sn, Cr, Ni, Pb, Cu, Cd, Zn, Co,
and Fe) had deposition profiles similar to that of the charcoal
particles (Figure 7.3). The recent decrease for these species is
probably related to the improved retention of fly ash released from
coal-and-oil burning facilities through the installation of control
devices. Other studies have indicated that there were lower levels of
atmospheric particulates in the atmosphere around Lake Michigan
beginning in the late 1960s.

Thus we can in principle derive a record of metals in the
atmosphere above Lake Michigan for the time period spanning
approximately the last century. With data on current atmospheric
concentrations and present-day sedimentary concentrations, we can
derive past atmospheric concentrations from sediment levels. Even
though different quantities of metals may come from oil and coal
burning, emissions from coke ovens, automobile exhausts, iron and
steel plants, and cement production, their overall atmospheric
concentrations are the important data for environmental regulators.

By correlating changes in fossil remains with pollutant deposition
records, insights into the ecological effects of pollutants can often
be obtained. Because long-term records are obtainable, creeping
changes can be separated from natural fluctuations and cycles.
Changes in the relative abundance of species, extinction of sensitive
forms, changes in preserved pigments, and changes in incidence of
deformities have all been used as criteria for assessing effects.
Paleolimnological techniques have been applied to the problems of
eutrophication, acidification, and general environmental contamination
(for example, Bradbury and Waddington 1973; Bradbury 1975; Gorham and
Sanger 1976; Davis and Berge 1980; Davis et al. 1980; Norton and Hess
1980; Strand 1980; Warwick 1980a,b). The increased use of such
techniques could partially overcome interpretive problems resulting
from inadequate long-term baseline data.

MONITORING LARGE-SCALE POLLUTION

No area of the United States or indeed of our planet is free from
the effects of air pollution. Acid precipitation is slowly eroding
the environmental quality of remote wilderness areas in northeastern

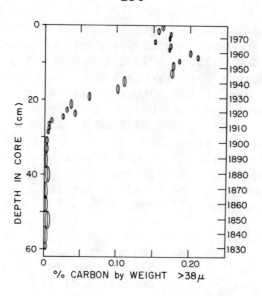

FIGURE 7.2   Charcoal (elemental carbon) concentration as a function of depth in the Lake Michigan core. SOURCE: Goldberg et al. (1981). Reprinted with permission from *Environmental Science and Technology*. Copyright © 1981 by the American Chemical Society.

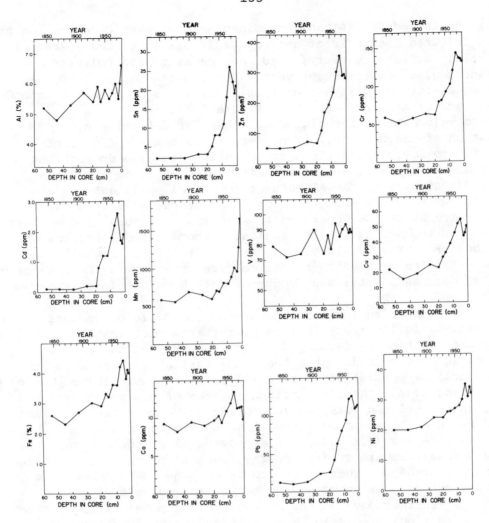

FIGURE 7.3   Metal concentrations (by dry weight) as a function of depth in the Lake Michigan core. SOURCE: Goldberg et al. (1981). Reprinted with permission from *Environmental Science and Technology.* Copyright © 1981 by the American Chemical Society.

Canada, the United States, and Scandinavia; atmospheric deposition has also contributed chemicals of anthropogenic origin to the icecap of the Arctic. Sulfate, mercury, lead, and cadmium have polluted Greenland's glaciers in recent years (Boutron and Delmas 1980), and vanadium, a micro-constituent of fossil fuel, has been traced into remote areas of Alaska (Kerr 1979, Figure 7.4).

The development of techniques suitable for detecting slow degradation of the biosphere is hindered by a number of current scientific policies and attitudes. Granting agencies do not fund long-term studies; most grants are limited to 1, 2, or at most 5 years. But studies of the effects of low-level pollutants on ecosystems must be designed to go on for decades, so that slow, long-term trends can be separated from seasonal, annual, and multiyear fluctuations in both natural ecological phenomena and emissions of air pollution. For example, large seasonal fluctuations in the hydrogen ion concentration of Swedish rivers obscured the slow overall tendency for pH to decrease, until many years of data were combined (Figure 7.5).

Because monitoring must now operate on the scale of decades rather than years, it is important that an appropriate coordination and funding organization be developed. Some funding for the initial phases of long-term studies has been made available recently by the National Science Foundation (Botkin 1977, 1978; Loucks 1979), but much remains to be done. The international system of biosphere reserves (Risser and Cornelison 1979) could provide suitable sites for baseline monitoring, and the proposed national network of experimental ecological reserves (TIE 1977) could provide sites for appropriate, long-term dose/response studies on the level of organisms, ecosystems, and whole watersheds. These ought to be combined with short-term laboratory experiments on both single-species and multi-species microcosms. Further development of theoretical models will also be needed, to test our predictive capabilities and to generate new insights into the behavior of species and ecosystems under pollution stress (West et al. 1980, Shugart et al. 1980). A recent National Academy of Sciences report on ecotoxicology reviews the techniques available to test the effects of chemicals upon ecosystems (NRC 1981).

Changes in scientific attitudes and educational systems will also be required. The deemphasizing of taxonomic work in North American environmental science for the past few decades has left us with a dearth of first-rate taxonomists and few institutions that offer competent taxonomic training. Sound taxonomy is a key ingredient in any broad-scale, long-term analysis of an ecosystem, or in the paleoecological methods required to interpret past pollution records. Likewise, environmental monitoring has for the most part become the piecemeal application of techniques designed for other purposes. Assessment of environmental pollutants and their effects must be transformed into a science in its own right, dedicated to the development and application of methods for detecting long-term, chronic effects on organisms and ecosystems.

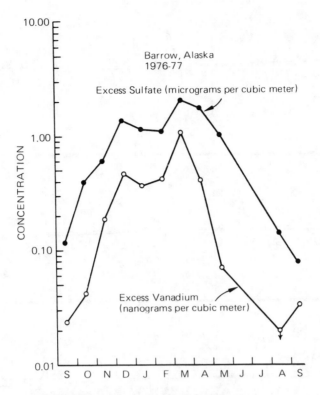

FIGURE 7.4 Monthly mean concentrations at
Barrow, Alaska, of atmospheric sulfate and vanadium that could not be attributed to natural
processes. Such excess vanadium is generally considered to result from the burning of heavy industrial fuel oils. SOURCE: Kerr (1979). Reprinted
with permission from *Science* 205:290-293. Copyright © 1979 by the American Association for the
Advancement of Science.

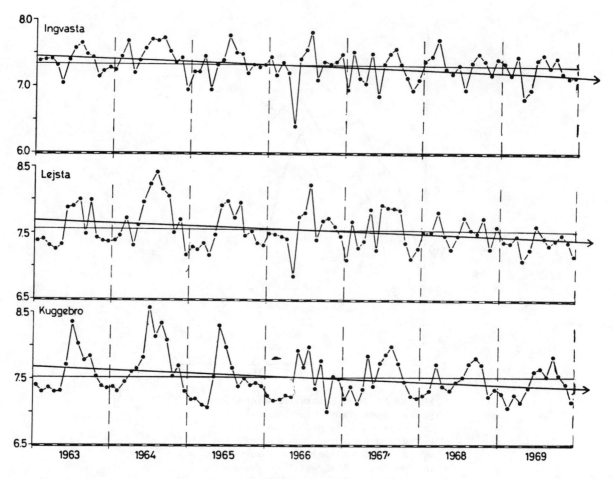

FIGURE 7.5   Changes in pH values of a Scandinavian river. The data points, from three stations in the Savjann River, show the yearly periodic swings, and the arrow shows the average pH value. SOURCE: Odén and Ahl (1970).

SUMMARY

The tremendous growth in the past few decades of our knowledge of atmospheric pollutants and their widespread dispersion has not been matched by an increase in our knowledge of their biological and ecological effects. Among the possible major effects envisaged are the climatic greenhouse effect, the serious perturbation of global sulfur and nitrogen cycles, the global dispersion of toxic trace metals, and the biotic impoverishment affecting the structure and function of major ecosystems. The study of atmospheric pollutants is very complex, because the substances must be followed from their sources of emission through atmospheric and ecosystem pathways--along which major transformations can occur--to biotic and abiotic receptors, which fluctuate naturally in abundance and distribution, and which are affected in extremely diverse ways. Linkages of pollutants are important for predictive purposes, and may occur at all points from the sources of initial emissions to the ultimate receptors where their effects are manifested.

Maximum pollution effect is expected where receptors are highly sensitive and ecologically important, and where sources and receptors are close together. (If the pollutant has a long atmospheric residence time such closeness is unnecessary.) For the prediction of effect it is important to know the distribution and abundance of sensitive receptors, the distribution and size of emission sources, and the properties of the pollutant that govern residence times in various ecosystem compartments. Sensitivity of organisms and ecosystems to pollutant stress may be caused by a wide variety of receptor properties, and may vary with the nature of the pollutant and the stage of development of the organism or ecosystem. Few generalizations are possible in this regard.

Stratigraphic studies of pollutant accumulation and changes in fossils in sediments or long-term deposits of several types can provide a useful guide to the history of a pollution problem.

Atmospheric pollution is now ubiquitous, but policies for investigating and dealing with it are not well developed. Evaluation of the long-term ecological effects of chronic environmental pollution are vital, but appropriate organizations to accomplish such evaluations are lacking, as are suitably trained personnel. There is an urgent need for new institutional arrangements and for training programs to establish ecotoxicology--the ecological assessment of environmental pollutants--as a new and independent scientific discipline.

8. ACID PRECIPITATION

The deposition of acid from the atmosphere was recognized in President Carter's second environmental message to the U.S. Congress on August 2, 1979, as "one of the most serious global pollution problems associated with fossil fuel combustion," rivaled only by the buildup of carbon dioxide in the atmosphere. The topic, generally known as "acid rain," has been of much international concern because acids are deposited far from the sources of their precursors.

Acid precipitation was the subject of a major international symposium in 1975, when research on many aspects of the topic was in its infancy. Since that time, many studies have been completed, including an enormous amount of work in Scandinavia. Recent symposia (Drablös and Tollan 1980, Hutchinson and Havas 1980, Shriner et al. 1980) treat many aspects of the problem in much more detail than previously, and some long-term studies of effects allow a much more conclusive summary of the problem. We can thus learn much from a case study of acid precipitation about the problems likely to be encountered with other atmospheric pollutants.

CAUSES OF ACID PRECIPITATION

It is thought that the pH of "pure" rain is controlled by the weak acid, carbonic acid (H_2CO_3), resulting from atmospheric CO_2 in solution. The resulting pH would be near 5.6. When alkaline dust and ocean sea spray are taken into account, the pH of precipitation must be higher than pH 5.6, probably around pH 7.0 (Odén 1976). But this theory is no longer testable, because global pollution of the atmosphere with sulfur compounds from fossil fuel combustion and the smelting of metalliferous sulfide ores has existed for decades, if not centuries, as shown by analyses of polar icecaps (Figure 8.1). It seems likely that under unpolluted conditions, small releases of sulfur and nitrogen oxides from volcanic acitivity and microbial metabolism to the biosphere would cause slightly lower natural pH levels than the theoretical level in geologically unbuffered areas, and that wind-carried calcareous dust would cause pH levels higher

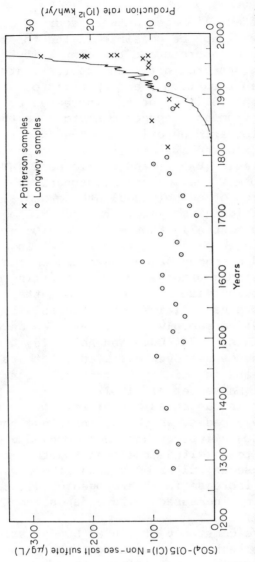

FIGURE 8.1 Sulfate concentrations on a sea-salt-free basis in northwestern Greenland glacier samples. The curve is the world production of thermal energy from coal, lignite, and crude oil (King 1969). SOURCE: Koide and Goldberg (1971). Reprinted with permission from the *Journal of Geophysical Research* 76:6589-6596. Copyright © 1971 by the American Geophysical Union.

than 5.6 in areas where calcareous rocks and soils are plentiful. It is now clear that precipitation is far more acid than theoretically pure rain in regions to which prevailing winds carry oxides of sulfur and nitrogen from heavily industrialized areas (Odén 1968, Dovland and Semb 1980).

Direct cause-and-effect linkages between sources of acid and effects on ecosystems will not be possible in the foreseeable future, owing to the remoteness of sources and the complexity of the interaction among emissions from different sources, atmospheric transport, chemical transformations, and specific orographic and geological settings (Table 8.1). But the increased emission of sulfur and nitrogen compounds from anthropogenic sources is the only plausible explanation for acid deposition. The oxides of these elements appear to be oxidized further in the atmosphere to form the strong acids, sulfuric acid (H_2SO_4) and nitric acid (HNO_3), which contaminate wet precipitation and atmospheric aerosols (Figure 8.2). The dry deposition of ammonium sulfate aerosols, which are then converted to sulfuric acid in ecosystems through biological processes, may also contribute substantially to the problem (Odén 1976).

While scientists are in general agreement that industrial emissions of sulfur and nitrogen oxides have caused the contamination of precipitation with strong mineral acids, the timing of the increase and the present rate of increase in acidity are matters of some dispute. Several authors have claimed that the acidity of precipitation has increased rapidly in the past few decades (for example, Cogbill and Likens 1974, Odén and Ahl 1970, Dickson 1975; see Figure 8.3). This evidence has been disputed by others, who claim that the apparent increase in acidity of precipitation is due to methodological changes (Hansen et al. 1981).

A number of changes in the emissions of acid precursors have taken place over the past few decades, which may influence the acidity of precipitation. While SO_2 emissions have not changed greatly for several decades, owing to a switch from coal to other fuels and to increased control of gaseous sulfur emissions, there was unquestionably a great increase in anthropogenic sulfur emissions in the 20th century, causing increased sulfate deposition in remote regions (see Figure 8.1).

Despite the relative constancy of annual SO_2 emissions during the past century, three technological changes may cause the emissions to produce acid precipitation more efficiently.

First, the height at which gases are injected into the atmosphere has increased nearly threefold (Fig. 8.4), causing SO_2 to be transported farther and to remain in the atmosphere longer, increasing the probability of oxidation to sulfuric acid. Second, recent controls of particulate emissions have reduced the amount of alkaline fly ash discharged from smokestacks, and it is conceivable that in the past such material partially neutralized acid emissions. Third, there has been a gradual change from seasonal to year-round emission. Less coal is used for space heating and more coal is used for generating electricity. Also the demand for electricity during the summer months

TABLE 8.1 Factors affecting the vulnerability of an ecosystem
to acid rain

A. Anthropogenic

    1. Spatial and temporal patterns of urban/industrial development
    2. Kinds and amounts of energy resources in use
    3. Controls on atmospheric emissions
    4. Degree of agricultural activity (cultivation, liming,
        fertilization)

B. Geologic

    1. Nature of bedrock, as regards both basic minerals and
        acid-soluble toxic metals
    2. Patterns of glaciation
    3. Depth, texture, mineralogy, and organic content of soil

C. Climatic

    1. Amount of precipitation
    2. Atmospheric humidity, as it affects gas absorption and
        particle collision
    3. Direction and speed of winds and air-mass movements
    4. Temperature, especially as it affects the proportions of rain
        and snow, and rates of chemical reaction in the atmosphere
    5. Ratio of precipitation to evaporation, as it affects leaching
        and the residence time of water in lakes

D. Topographic

    1. Altitude, as it influences soil depth, precipitation, etc.
    2. Order of streams and lakes in the hydrologic network
    3. Lake depth and ratio of watershed area to lake area,
        controlling residence time of water

E. Biotic

    1. Height, type, and duration of leaf canopy
    2. Magnitude of transpiration
    3. Sensitivity of critical species, including the microbes
        mediating biogeochemical cycles

F. Natural, episodic

    1. Volcanoes, producing locally acid rain
    2. Fires in deposits of fossil fuel such as coal or lignite
    3. Forest fires, entraining alkaline particulates into the
        atmosphere
    4. Dust storms, entraining alkaline soil particles into the
        atmosphere

144

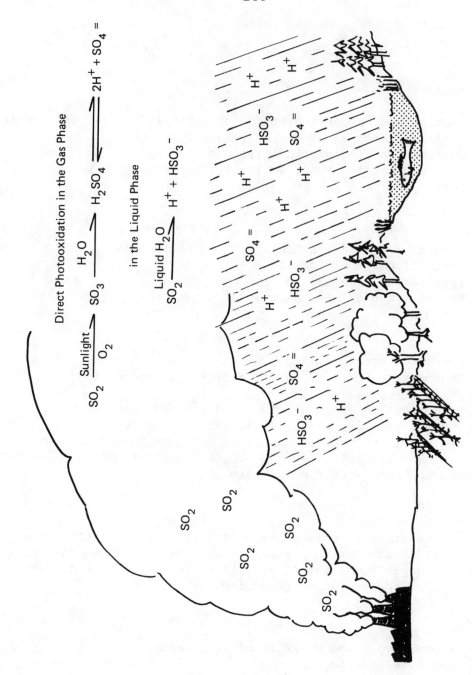

Direct Photooxidation in the Gas Phase

$$SO_2 \xrightarrow[O_2]{Sunlight} SO_3 \xrightarrow{H_2O} H_2SO_4 \rightleftharpoons 2H^+ + SO_4^=$$

in the Liquid Phase

$$SO_2 \xrightarrow{Liquid\ H_2O} H^+ + HSO_3^-$$

FIGURE 8.2 The formation of sulfuric and sulfurous acids from sulfur oxide pollutants.

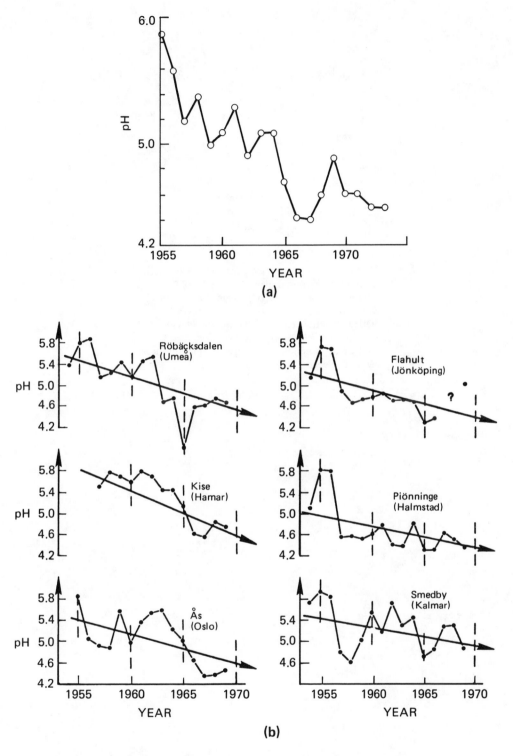

FIGURE 8.3 The pH levels of precipitation in Scandinavia, 1955-1975. SOURCES:
(a) Dickson (1975); (b) Odén and Ahl (1970).

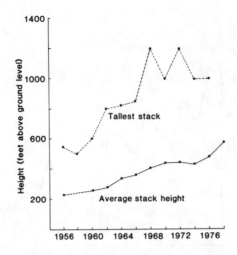

FIGURE 8.4   Average stack height and tallest stack
reported among power plants burning fossil fuels
(bituminous coal, lignite, oil) included in biannual
design surveys of new power plants, 1956-1978.
SOURCE: Patrick et al. (1981). Reprinted with permission from *Science* 211:446-448. Copyright ©
1981 by the American Association for the Advancement of Science.

has grown with the increased use of air conditioning. The high temperatures and humidities in summer may result in more efficient oxidation of SO_2 emissions to sulfuric acid.

In addition to the technological changes in SO_2 emission, there has been an increase in emission of nitrogen oxides for the past few decades (see Table 4.2). Nitrogen oxides are emitted from a wide variety of sources, with some injected high into the atmosphere while others are ejected and dispersed at ground level through motor vehicle use. In the absence of control technology for nitrogen oxides, their emissions will exceed emission of sulfur oxides by the turn of the century.

We stress that emission of nitrogen and sulfur oxides and the consequent acid precipitation are broad regional rather than global problems. When natural and anthropogenic emissions are compared on a regional basis, it is clear that man's acitivities completely overwhelm natural sources of SO_2 and NO_x, (see Table 4.1), even though the magnitude of anthropogenic emissions of these oxides may seem unimportant when compared with natural emissions on a global scale.

The observed recent increases in lake acidity could have resulted either from a rapid increase in acid precipitation in recent time or from long-term, constant acid precipitation over several decades duration. It is difficult to differentiate between these two possible patterns because of the nature of the bicarbonate buffering curve. As a solution of bicarbonate--such as a lake--is titrated by a constant addition of strong acid there is little resulting change in pH until 80 to 90 percent of the bicarbonate has been consumed according to the reaction:

$$H^+ + HCO_3^- \longrightarrow H_2CO_3 \longrightarrow H_2O + CO_2$$

Once the bicarbonate has been converted into carbon dioxide and lost to the atmosphere, the acid (hydrogen ions) accumulates and the pH decreases rapidly (Figure 8.5). The relatively sudden drop in pH to an acid condition is the same regardless whether the titration (the addition of acid) occurred at a fast or slow rate. The lake will become acidic so long as the rate at which acid is added to the lake exceeds the rate at which geochemical weathering processes replace the bicarbonate.

The theory that the acidification observed in poorly buffered fresh waters was due to changing land-use patterns (Rosenqvist 1978a,b) has now been discounted as an explanation for the widespread effects observed, particularly in remote areas. Detailed study over several years of watersheds in Norway, some with changing land-use patterns and some without, has shown that, on the average, both are acidified at equal rates (Drablös and Sevaldrud 1980, Drablös et al. 1980). Moreoever, studies of lakes in North America in areas where land-use patterns have never changed have also shown substantial increases in hydrogen ion or losses in buffering capacity (Dillon et al. 1978, Watt et al. 1979).

Along with the hydrogen ions supplied by transformation of atmospheric sulfur dioxide and oxides of nitrogen to their soluble,

148

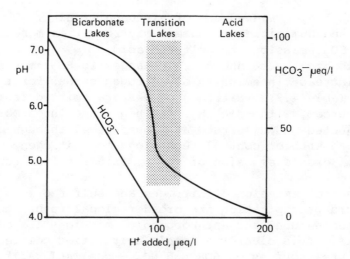

FIGURE 8.5   Titration curve for bicarbonate solution at a concentration of 100 μeq/l, illustrating the acidification process. SOURCE: Henriksen (1980).

acid forms, there is a potential for ecosystem acidification by the
nitrification of ammonia, from atmospheric precipitation or from the
decomposition of dead organic matter. Two equivalents of hydrogen ion
are generated for each equivalent of ammonium ion transformed to NO_3
(Reuss 1975a), but one of those equivalents may be consumed upon
either uptake or denitrification by components of the biota. If all
of the ammonium and nitrate ions are transformed and/or utilized by
organisms, the net potential for ecosystem acidification will be
measured by the difference of ammonia and nitrate in equivalents.
Preliminary data show very high concentrations of ammonia in
precipitation over the central United States, possibly as a result of
crop fertilization and livestock culture (Figure 8.6). Odén (1976)
has compared the trends in southern Sweden of anthropogenic
acidification directly by mineral acids and indirectly by biological
processes, and the normal background of biological acidification by
nitrogen transformation (Figure 8.7). Mayer (1979) discussed the
conditions under which natural acidification may be most significant.

Extent of the Problem

Large areas of the earth's surface consist of poorly buffered
geologic materials. Where such areas occur within several hundred
kilometers of sources of atmospheric emissions that are acid
precursors, detectable acidification of at least freshwater ecosystems
may be expected to occur. In the United States such areas occur
largely in the eastern part of the country (Cogbill and Likens 1974).
In Canada, there are roughly 2 million square kilometers of
acid-sensitive terrain--most of the eastern half of the country.
Similar large geologic areas occur in Scandanivia, Scotland, and the
northern part of the Soviet Union. A high proportion of the world's
freshwaters occur in such terrain (between 50 and 80 percent depending
on estimates); thus, large-scale degradation becomes an important
concern.

EFFECTS OF ACID ON THE BIOSPHERE

Aquatic Ecosystems

Widespread and pronounced effects of acid precipitation have been
recorded in poorly buffered aquatic ecosystems. The low chemical
capacity of such "soft" waters to buffer against increasing
hydrogen-ion supply causes the water to be in precarious pH balance
even under natural conditions. Indeed, many natural lakes may become
more acidic over hundreds or thousands of years owing to drainage from
acid peat bogs. The bog moss, Sphagnum, generates in its cell walls
polyuronic acids (Clymo 1964), whose hydrogen ions are exchanged for
metal cations from precipitation (Gorham and Cragg 1960). The
hydrogen ions that are released in this way acidify the bog waters.

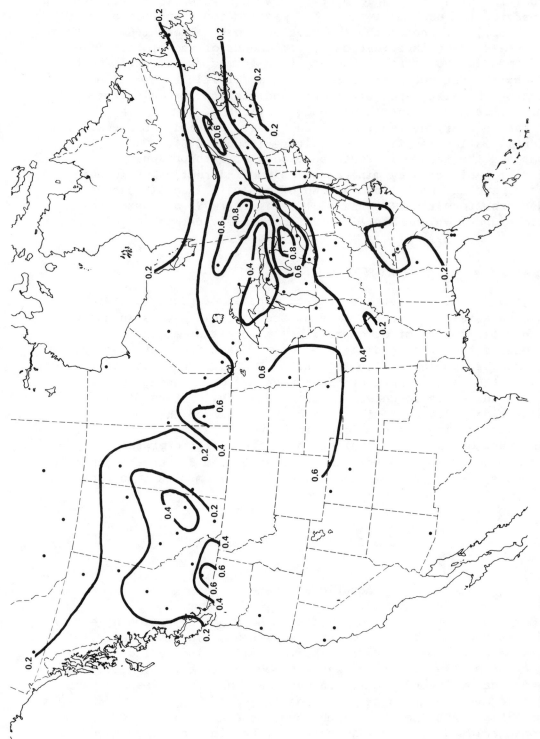

FIGURE 8.6 Mean annual concentrations of ammonia in precipitation in the United States and Canada, 1979-1980. The isopleths show ammonia concentrations in milligrams per liter and are based on data from the National Atmospheric Deposition Program (NADP) in the United States and the Canadian Network for Sampling Precipitation (CANSAP) collected from April 1979 through March 1980. NADP data points are averaged from 16 or more weekly samples, and CANSAP sites of 4 or more monthly samples were included. SOURCE: W. W. Knapp, Department of Agronomy, Cornell University, Ithaca, New York, personal communication, 1981.

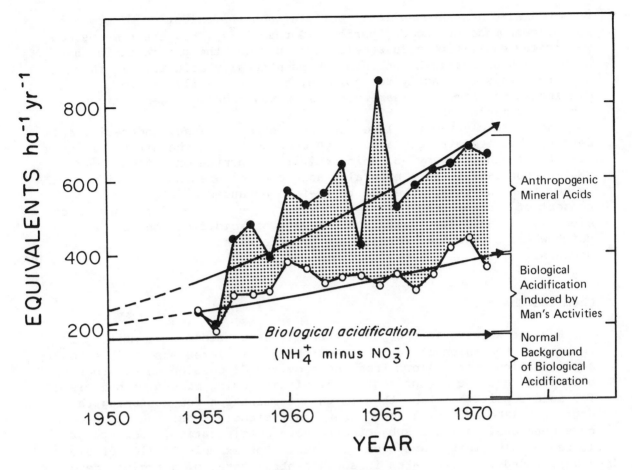

FIGURE 8.7   Acidification due to mineral acids (computed as excess acids) and to biological processes. The computations are based on data on the southwestern part of Scandinavia from the International Meteorological Institute. SOURCE: Odén (1980).

Bogs may also generate sulfuric acid through the oxidation of organic
sulfur compounds during dry periods (Gorham 1967). By these means bog
waters may exhibit pH values below 4 even where the precipitation has
a pH of 5.0 or more. Lakes and streams naturally acidified in this
way are always distinctly tea-colored, and can readily can be
distinguished from the clear-water lakes now undergoing acidification
because of man's activities.

There are clear-water lakes in areas such as Japan, Indonesia, and
Germany that are strongly acidic (pH down to 0.8). The strong mineral
acidity is generated by volcanic activity or pyrite oxidation
(Hutchinson 1957). Such naturally acidic lakes are the exception
rather than the rule. The 10- to 40-fold enhanced hydrogen-ion
content of current precipitation, however, has increased the number of
clear-water, acidified lakes and the rate of acidification to the
point where substantial changes are notable within a decade or
decades.

Chemical Changes

Routine surveys in the past generally have used measurement
techniques too insensitive to detect changes in lakes susceptible to
acidification. It follows from the previous discussion on bicarbonate
buffering that the measurement of alkalinity using modern methods is a
far more sensitive method than is pH determination for measuring the
degree of lake acidification. Where sufficiently sensitive techniques
have been used in lakes subjected to acid precipitation, spectacular
losses of alkalinity have been recorded. For example, Dillon et al.
(1978) found that two lakes in south central Ontario, a region where
the average pH of precipitation is 3.95 to 4.38, had lost over 50
percent of their alkalinity in a period of 5 to 10 years. Seasonal
data are necessary in such studies to separate long-term trends from
annual variation, because alkalinity in soft water systems is also
greatly affected by seasonal cycles of plant production.

More typically, long-term measures of alkalinity are not available
and the effects of acid precipitation must be interpreted through
changes in pH. Watt et al. (1979), Thompson et al. (1980), the
Ontario Ministry of Environment (1978), and Harvey (1980) have shown
moderate to substantial decreases in pH over periods of a few years to
two decades. In the latter two studies, rates of pH change (once
again in south central Ontario) were often 0.1 pH units per year or
more, indicating the total exhaustion of bicarbonate buffering in the
lakes. Major damage to lakes may occur well before such dramatic
reduction in pH takes place. For example, in an experimental
acidification of a small lake, Schindler et al. (1980) found that 70%
of the lake's alkalinity had been depleted before pH values decreased
detectably below normal.

Some studies show a decline in the pH of lakes over the past two
decades, although early surveys were done colorimetrically and then
later on by electrode so that the magnitude of observed changes may
not be entirely accurate. Examples are those of Schofield (1976) for
the Adirondack area of the United States and Wright et al. (1980) for
Norway.

Another chemical index of acidification is found in the ionic ratios of lakes. When the contribution of sea salts has been removed by indexing other ions to the chloride ion, most lake-water chemistry is dominated by calcium or calcium and magnesium, and the anions are dominated by bicarbonate--a reflection of the fact that the weathering products of rocks and soils which control the ionic composition of lakes are similar throughout most of the northern hemisphere (Rodhe 1949).

As sulfuric acid is supplied by precipitation, the bicarbonate is consumed according to the above equation, being replaced in the ionic balance by the sulfate ion from precipitation. As a result, as lakes are acidified, there is a shift from a solution dominated by Ca^{++} and HCO_3^- to one dominated by Ca^{++} and $SO_4^=$. The relative increase in $SO_4^=$ and H^+ or the relative decrease in HCO_3^- with respect to Ca^{++} or Ca^{++} plus Mg^{++} has been the basis for a number of models to predict the degree of acidification. For example, Henriksen (1979) uses a plot of pH versus Ca^{++} (Figure 8.8) while Almer et al. (1978) employ a plot of HCO_3^- (or alkalinity) versus Ca^{++} plus Mg^{++} (Figure 8.9). While the ratios may be affected to some degree by the amount of calcium and magnesium weathered from terrestrial catchments, there is little doubt that the index is a reliable indicator of damage due to acidification where noncalcareous bedrock and soils predominate. The possible effect of increased cation leaching on the magnitude of the ratio is addressed by Henriksen (1980).

Scandinavian and North American studies appear to agree on one point: acidification of sensitive waters is detectable within one to two decades where pH values of precipitation are less than 4.6--a 10-fold increase in acidity over the theoretical "pure" rain pH value of 5.6 (Henriksen 1979, 1980; Watt et al. 1979; Thompson et al. 1980). In areas where precipitation has pH values of 4.0 to 4.3, degradation is much more rapid (Dillon et al. 1978, Keller et al. 1980). Changes in the pH or alkalinity of water bodies of small areas of Belgium, the Netherlands, and Denmark also have been recorded (Vangenechten and Vanderborght 1980, Van Dam et al. 1980, Rebsdorf 1980).

Metal concentrations also tend to increase in acidified lakes. While this may be due in part to the co-deposition of hydrogen ion and trace metals emitted from anthropogenic sources (discussed in Chapter 4), ion exchange from acidified soils and lake sediments are more important sources in many cases.

Most notable is the correlation between hydrogen ion and aluminum in lakes from a number of areas (for example, Fig. 8.10). Aluminum is abundant in both soils and sediments. In terrestrial catchments, it exchanges for hydrogen ion, and when the pH of groundwater or surface flow is 5 or less, high concentrations of the element may be carried to lakes and streams (Johnson 1979, Cronan and Schofield 1979, Johannessen 1980, Hermann and Baron 1980).

The chemistry of aluminum in lake water is a complex function of pH, sulfate, and nitrate concentrations. It is reviewed by Almer et al. (1978). In general, at pH values above 5.5, aluminum forms

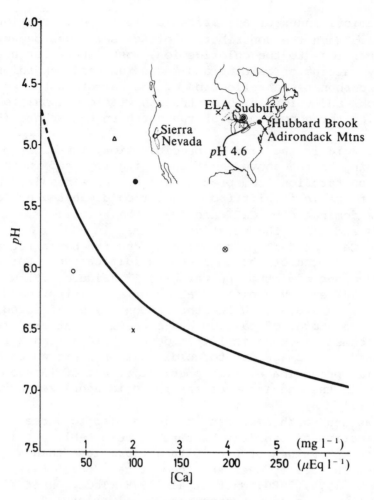

○ average for 170 lakes in the Sierra Nevada, California

× average for 109 lakes in the Experimental Lakes area, northwestern Ontario

● average for 216 lakes in the Adirondack Mountains, New York

△ average for 11 years at Hubbard Brook, New Hampshire
(Schofield, personal communication)

⊗ average for 178 lakes in the vicinity of Sudbury, Ontario

FIGURE 8.8   Average pH levels and calcium concentrations in
regions of North America. Waters in areas receiving acid precipitation lie well above the solid curve, whereas those in other areas
lie below. Inset shows locations of these areas. Areas east of the
isoline receive precipitation more acidic than pH 4.6. SOURCE:
Henriksen (1979). Reprinted with permission from *Nature* 278:
542-545. Copyright © 1979 Macmillan Journals Limited.

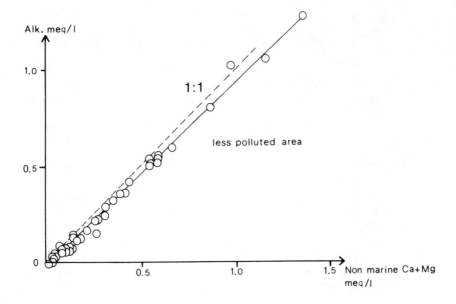

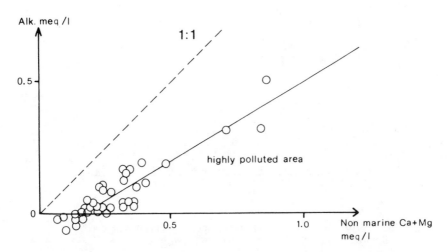

FIGURE 8.9   Alkalinities and contents of calcium and magnesium of nonmarine origin in lakes in two regions in Sweden with different sulfur loads. SOURCE: Almer et al. (1978).

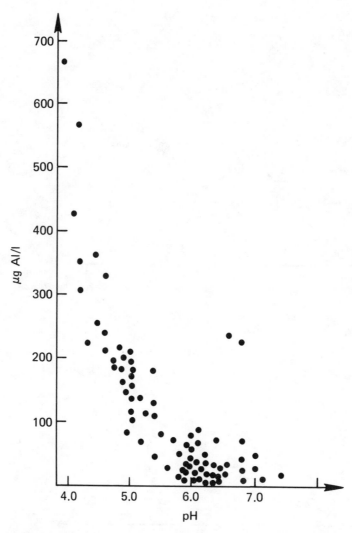

FIGURE 8.10 The pH values and aluminum contents in lakes on the
Swedish west coast, 1976. SOURCE: Dickson (1980).

complexes that will tend to precipitate from solution. Below pH 5, several weak acid species predominate. Toxicity is high at low pH, as discussed later.

Aluminum has a pronounced effect on other chemical cycles. For example, it is known to precipitate the humates that cause the dissolved color of lakes, leading to increased transparency. The increased light penetration causes higher phytoplankton production deep in the water column (Almer et al. 1978, Schindler 1980b), and the thermocline may deepen due to greater penetration of solar energy.

Dickson (1980) has shown that increased aluminum concentrations precipitate phosphorus from lake water, particularly in the pH range 4.5 to 6.0 (Figure 8.11). Because phosphorus is the key element controlling the productivity of many lakes, this secondary effect of acidification may be responsible for the apparent "oligotrophication" which has been observed (Grahn et al. 1974). The co-precipitation with aluminum appears to more than counteract the increased deposition of phosphorus from polluted precipitation (Dickson 1980).

Hultberg and Wenblad (1980) found high concentrations of aluminum and manganese in acidified ground water. Their study provides an excellent example of how a problem may develop quickly as acid deposition continues. High sulfates reaching an aquifer from surface deposition of acids caused no acidification for many years, because a high water table in the area allowed microbial reduction of the sulfate to sulfide, which was then precipitated as pyrite by combination with dissolved iron. After two dry years, however, during which the groundwater table lowered, the accumulation of sulfide from several years was reoxidized rapidly to H_2SO_4, causing highly acid groundwater, which leached high concentrations of several cations from surrounding soils including the above two metals.

Manganese, zinc, nickel, lead, and cadmium also appear to be washed into lakes and streams from acidifying terrestrial systems (Figure 8.12, Dickson 1980; Hutchinson et al. 1978; Havas 1980; Troutman and Peters 1980) and mobilized from lake sediments (Schindler et al. 1980b). Schindler and his colleagues found concentrations of zinc known to be toxic to aquatic animals to be supplied from sediments into overlying water at pH 5.1, while Havas (1980) found a similar situation for both zinc and nickel.

At low pH, methylated forms of mercury appear to occur in the monomethyl form (Figure 8.13), leading to its more rapid accumulation in fish. High mercury concentrations in fish from acidified waters have been reported from the United States, Norway, and Sweden. Jackson et al. (1980) found that adsorption of the isotope mercury-203 to organic materials in sediments at pH 5.1 was much lower than at near-neutral pH. Ionic mercury also appears to be more efficiently scavenged from the atmosphere than elemental mercury by aerosols, clouds, and rain with low pH levels. Thus acid precipitation, coupled with the enhanced emission of mercury to the atmosphere described in earlier chapters may cause serious problems.

While hydrogen-ion activity affects the speciation of a host of other trace metals, there is as yet no evidence for their increased mobilization from lake sediments at low pH. Schindler et al. (1980b)

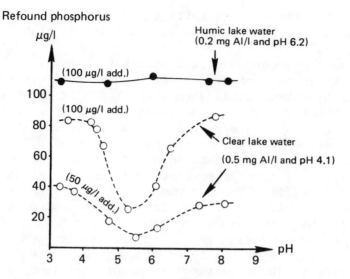

FIGURE 8.11 Phosphorus recovered from the supernatant after experimental additions of orthophosphate to lake water samples. Fifty and 100 μg of orthophosphate were added to one-liter samples from Lake Horsikan (clear lake water with an initial pH of 4.1 and containing 0.5 mg aluminum per liter), and 100 μg of orthophosphate was added to one-liter samples of humic lake water (with an initial pH of 6.2 and containing 0.2 mg aluminum per liter that is mainly complexed with dissolved humic substances). The pH of the water was adjusted to different values, and the water samples were stored for 5 days before analysis. SOURCE: Dickson (1980).

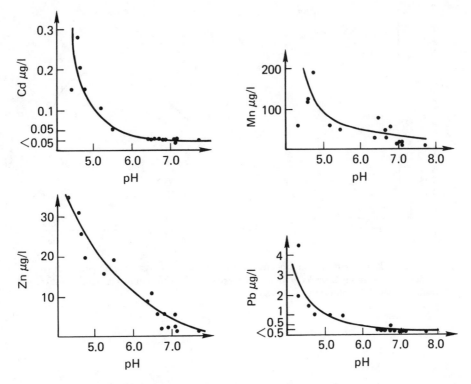

FIGURE 8.12 Metals in 16 lakes on the Swedish west coast with similar metal deposition but with different pH, December 1978. SOURCE: Dickson (1980).

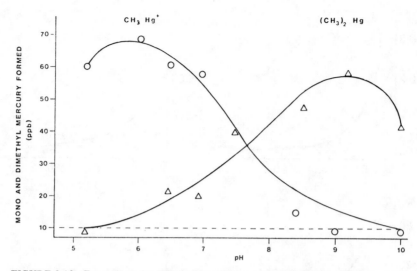

FIGURE 8.13 Formation of mono- and dimethyl mercury in organic sediments at different pH levels during 2 weeks, with a total mercury concentration of 100 ppm in substrate. SOURCE: Tomlinson et al. (1980).

found that detectable copper, cadmium, cobalt, and lead were not released by sediments at pH values down to 5.1, while zinc, manganese, aluminum, and iron were mobilized from sediments at this pH. In general, low pH favors transformation of metals into more toxic ionic forms.

While theory predicts that acidification should negatively affect microbial denitrification, ammonification, and nitrification, as observed in terrestrial ecosystems, no evidence on these effects is currently available.

Silica has been thought to be unaffected over a pH range of 2 to 8 (Birkeland 1974, Driscoll 1980). Concentrations in Swedish lakes, however, are lower at lower pH levels (Figure 8.14). This fact is currently inexplicable, particularly in view of the observed decrease in diatoms as pH decreases.

Biological Changes

As a result of acidification a number of organisms have been reduced or eliminated over significant parts of their ranges. A considerable change in algal species results from acidification and the changes associated with it. Some species of diatoms and chrysophyceans are eliminated in both plankton and periphyton communities, and are replaced by chlorophytes or cyanophytes (Schindler 1980a, Müller 1980, Lazarek 1980). The total number of phytoplankton species is known to be lower in acid lakes (Raddum et al. 1980). In lakes where cyanophytes dominate instead of chrysophytes, acidification appears to cause a shift toward chlorophyte domination (Crisman et al. 1980, Yan and Stokes 1978). In some cases, the genus Mougeotia produces a filamentous mat on littoral sediments (Schindler 1980a, Almer et al. 1978) as a lake becomes more acid. In acid lakes in Ontario, the algae Zygnema and Zygogonium form benthic mats. Although lakes appear to become clearer as they are acidified, leading some authors to fear that they are becoming less productive (Grahn et al. 1974), it is still not known whether the increased transparency is due to loss of productivity or a change in the color of dissolved humic material. The phenomenon does not appear to be a direct result of acidification (Schindler 1980b), but may be due to the precipitation of both phosphorus and humates by aluminum leached from acid-stressed terrestrial materials, as illustrated in Figure 8.11.

Effects are severe on animals as well as plants. Studies of distribution show that numerous lacustrine molluscs and crustaceans are not found even at weakly acid pH values of 5.8 to 6.0 (Figure 8.15; J. Økland 1980, K. Økland 1980, Økland and Økland 1980). Experimental, whole-lake acidification studies have revealed similar tolerances for benthic Crustacea (Schindler 1980a). The crayfish Orconectes virilis rapidly loses its ability to recalcify after moulting as the pH drops from 6.0 to 5.5 (Malley 1980) and is probably thus rendered more vulnerable to predation and protozoan infection. In Scandinavia, the once-common crayfish Astacus astacus has become rare in lakes where the pH is below 6. The phyllopod Lepidurus

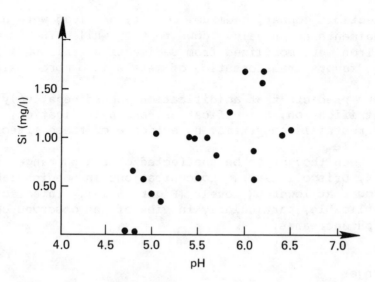

FIGURE 8.14 Concentration of silicon in relation to pH level in 20 Swedish west coast lakes, August 1978. SOURCE: Dickson (1980).

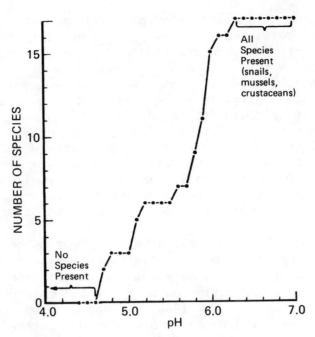

FIGURE 8.15 The pH tolerance limit for 17 widespread species of molluscs and crustaceans. SOURCE: Ökland and Ökland (1980).

arcticus is not found in water below pH 6.1 (Almer et al. 1978).
Other planktonic crustaceans such as Daphnia species and the fairy
shrimp, Branchinecta paludosa are also very sensitive to small
declines in pH below 5.5. Daphnia magna and D. middendorfiana are
susceptible to fungal infections at low pH (Havas 1980). The effects
of heavy metals are negligible at such high pH values, and thus the
direct toxicity of hydrogen ion is implicated.

Fish have been more extensively studied than other aquatic
organisms in acidified lakes. Fromm (1980) believes that impairment
of reproduction occurs at any pH below 6.5. Kennedy (1981) found high
incidences of embryonic abnormality and high embryonic mortality rates
of lake trout at pH 5.8. The reproductive failure appears to be due
to disruption of calcium metabolism and deposition of protein in the
oocyte (Fromm 1980).

Other stages of the life cycle usually appear to be less
sensitive. Adult fishes of most species survive at pH values of 5.0
to 5.5. Changes in ionic balance appear to result from exposure to
acid. Effects seem to be mitigated by increased calcium
concentrations (McDonald et al. 1980). Blood pH also decreases,
resulting in less capacity to carry oxygen. The fathead minnow,
Pimephales promelas, disappeared from an experimentally acidified lake
at pH 5.8 (Schindler 1980a). Spawning and egg production are known to
be affected at such pH values (Spry et al. 1981; Figure 8.16).

At slightly lower pH values, other fish populations begin to
decline. At pH values less than 5.0, most of the species of value to
sport or commercial fisheries have disappeared (Figure 8.17). In
Scandinavia, stocks of the sensitive roach, Rutilus rutilus, were
destroyed by acidification of some lakes as early as the 1920s. This
cyprinid and the minnow Phoxinus phoxinus require pH values greater
than 5.5 for successful reproduction (Almer et al. 1978). Salmonid
species react similarly to their North American counterparts (Figure
8.17).

The combination of aluminum and hydrogen ion is highly toxic to
fish at concentrations where neither is toxic alone. The fish kills
observed at spring melt in Scandinavia and the Adirondacks appear to
be due to a combination of high acidity from melting snow plus high
levels of aluminum leached from terrestrial soils (Cronan and
Schofield 1979, Driscoll 1980, Baker and Schofield 1980, Grahn 1980,
Muniz and Leivestad 1980). The effect appears to be due to a clogging
of the gill by irritation-induced mucus discharges, causing severe
respiratory stress. The loss of plasma sodium and chloride also
occurs (Muniz and Leivestad 1980, Rosseland 1980).

Studies of the effects of acidification on aquatic microbial
communities are scarce. In some cases, decreased decomposition of
organic matter has resulted from acidification (Traaen 1980), while in
others none has been observed (Schindler 1980a). It is possible that
these differences are due to differences in trace metal
concentrations. Sulfate reducers are stimulated by the increased
sulfate that accompanies sulfuric acid inputs, and under appropriate
environmental conditions they may partially counteract the effects of
acidification (Schindler et al. 1980a).

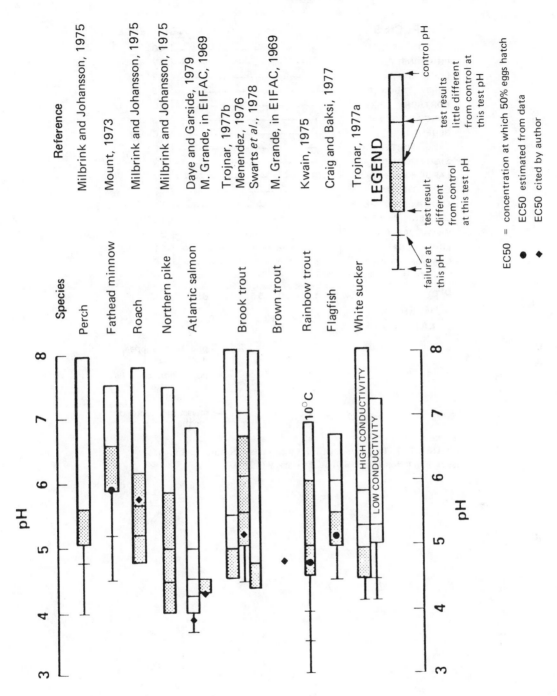

FIGURE 8.16 Egg hatchability of several fish species as a function of pH. SOURCE: Spry et al. (1981).

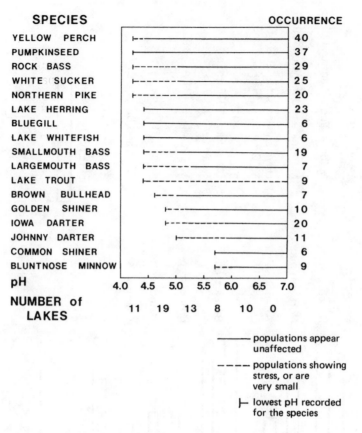

FIGURE 8.17 Occurrence of fish species in six or more La Cloche lakes, in relation to pH. SOURCE: Harvey (1980).

Some indirect effects of acidification on community structure have also been demonstrated. Many insect larvae survive well in the sediments of acidified lakes, especially chironomids. As predatory fishes disappear, more acid-resistant carnivorous invertebrates increase in numbers to fill the vacant trophic niche, including corixids, Chaoborus larvae, and dytiscid beetles (Henrikson et al. 1980). After the disappearance of Pimephales as mentioned above, the previously rare pearl dace, Semotilus margarita, increased rapidly to fill the vacant ecological niche.

Terrestrial Ecosystems

Vegetation

Adverse effects of acid precipitation on forests have not been proven. It is difficult to assess the effect of acid precipitation on forest yield against a background of yield differences caused by annual climatic variation. In addition, over much of the area subjected to acid precipitation, the soils are naturally acidic podsols with a vegetation adapted to these acidic conditions. Indeed, some experimental studies of Scandinavian forested areas show the additional sulfur and nitrogen supplied by acid precipitation to have a slight fertilizing effect in the short term, particularly in mature podzolic soils, which are capable of adsorbing large amounts of sulfate (Abrahamsen 1980, Tamm and Wiklander 1980, Tveite 1980). Long-term effects are uncertain, but the initial stimulation caused by adding nitrogen to a nitrogen-deficient forest may give way to deficiencies of other elements, such as Ca, K, Mg, P that are more rapidly leached away. As magnesium and other elements are mobilized at low pH and leached from the soil, long-term permanent damage to the ecosystem may result, causing chronic deficiencies of the elements in plants. Also increased mobilization of ions could result in toxic concentrations of aluminum and manganese. Seedling establishment may be affected. There is a paucity of studies of understory vegetation, although Horntvedt et al. (1980) mention reduced moss cover under forests artificially subjected to precipitation of pH 3, and Evans and Curry (1979) found that gametophyte fertilization in bracken fern was highly sensitive to precipitation of pH 3.4.

Effects on crops as a result of direct foliar damage have been reported, though effects are species specific and depend on environmental and physiological conditions. Some positive effects of acid precipitation on soybean yield have been observed (Irving and Miller 1980). However, Evans et al. (1980) found that simulated rain at pH 4 and below the caused number of seed pods per plant to be reduced and thus the seed mass of soybeans produced was significantly decreased. In a comprehensive review of the subject, Jacobson (1980) concluded that there was experimental evidence for damage to agricultural crops by strongly acid precipitation. Rains with a pH below 3.0 have been noted in many experiments to cause foliar lesions (Tables 8.2 and 8.3). Differences among studies may reflect differences in soil type or treatment technique.

TABLE 8.2   Results of Recent Experiments on Effects of Simulated Acidic Precipitation on Crops Grown Under Greenhouse Conditions

Laboratory	Crop	Effect	pH
Argonne National Laboratory	Soybean	Foliar symptoms	3.0
		No effect on growth	3.0
Boyce Thompson Institute	Lettuce	Increased growth	3.0 and 3.2
		Increased and decreased nutrient content	(dependent on sulfate and nitrate concentrations)
Brookhaven National Laboratory	Pinto bean	Reduced growth	2.5, 2.7, 2.9, 3.1
		Reduced yield	2.5, 2.7
	Soybean	Reduced growth	2.5, 2.7, 2.9, 3.1
		Reduced yield	2.5
		Increased yield	3.1
Cornell University	11 of 13[a]	Foliar symptoms	4.0
	13 of 13[a]	Foliar symptoms	3.0
	3 of 11[a]	Reduced growth (ht.)	3.0
	5 of 14[a]	Reduced growth (wt.)	3.0
	Tomato, cabbage, and pepper	Reduced yield and quality	3.0, 4.0
		Reduced growth and yield	3.0
Corvallis Environmental Research Laboratory	5 of 35[a]	Foliar symptoms	4.0
	28 of 35[a]	Foliar symptoms	3.5
	31 of 35[a]	Foliar symptoms	3.0
	5 of 28[a]	Decreased yield	3.0, 3.5, 4.0
	6 of 28[a]	Increased yield	3.0, 3.5, 4.0
Oak Ridge National Laboratory	Red kidney bean	Foliar symptoms	3.2
		Reduced growth	3.2, 4.0

[a]Number of species exhibiting effect out of total number exposed.

SOURCE: Jacobson (1980).

TABLE 8.3   Results of Recent Experiments on Effects of Simulated Acidic Precipitation on Field-Grown Crops

Laboratory	Crop	Effect	pH
Argonne National Laboratory (rain and simulated rain)	Soybean "Wells"	No effect on seed mass Increase in seed size No foliar symptoms or effects on growth	3.1
Boyce Thompson Institute (simulated rain only)	Soybean, "Beeson" and "Williams"	Decreased growth, yield, and seed quality (germination) Increased yield No foliar symptoms	2.8 (high ambient ozone) 2.8, 3.4 (low ozone) 2.8, 3.4, 4.0 (low ozone)
Brookhaven National Laboratory (rain and simulated rain)	Soybean, "Amsoy"	Decreased yield and quality (protein content), foliar symptoms	2.3, 2.7
Cornell University (rain and simulated rain)	Tomato, pepper, snapbean, cucumber	No effect on growth or yield, reduced quality	3.0
North Carolina State University (rain and simulated rain)	Soybean, "Davis"	Slight foliar injury No effect on growth or yield	2.8 2.8, 3.2, 4.0

SOURCE: After Jacobson (1980).

Plants differ greatly in their responses to pH (Russell 1973), some being adapted to calcareous soils (calcicoles) and others to acid soils (calcifuges). Still other plants tolerate a wide range of pH, and indeed species vary in their degree of tolerance in different parts of their range (Snaydon 1962). Many acid-tolerant species, moreover, are found in nature on acid soils because they can compete most successfully there; in the absence of competition they may grow as well or better on neutral soils. Likewise, some calcicoles can grow on acid soils in the absence of competition (Weaver and Clements 1938). For more than 50 years it has been generally accepted that the elevated aluminum and sometimes manganese concentrations in acid soils are key factors in determining survival of many species (Rorison 1980).

The effects of acidification upon biota are often extremely difficult to establish, owing to the diverse ways in which pH affects the availability of nutrients and toxicants, which may sometimes counterbalance one another. Even for a single nutrient element the situation may be highly complex, as shown in the hypothetical model suggested by Tamm (1976) for the effects of acidification upon the nitrogen cycle in a forest ecosystem (Figure 8.18). Abrahamsen and Dollard (1979) have recently reviewed the complex effects of acid rain on the nitrogen cycle. The potential effects of acid rain on foliar susceptibility of fungal attack and of roots to infection by fungal pathogens is an area of concern, but one lacking a detailed examination (Shriner and Cowling 1980). Most forest species require a symbiotic association with specific soil fungi (mycorrhiza) for adequate mineral uptake. This delicate association--particularly the infection stage--may be affected.

Soils

Acid rain has particularly severe effects upon the cycles of metallic toxicants and nutrients, because of the influence of hydrogen ions upon cation exchange and weathering and the consequences of mobilizing various elements.

An excellent example of such effects is provided by Gjessing et al. (1976), who observed that in small, undisturbed granitic watersheds, the sum of Ca + Mg + Al output was closely related to input of hydrogen ions. Metal output (O, as kg equivalents $km^{-2} yr^{-1}$) is correlated very highly (r = 0.99) with net retention and neutralization of hydrogen ions (R), the regression being given by O = 14.9 + 0.86 R. With high input of acid rain, metal output only slightly exceeded the retention of hydrogen ions. However, at low acid-rain input, the metal output is substantially in excess of the retention of hydrogen ions, because where rain is not strongly acid, natural inputs of hydrogen ions from such processes as decomposition and root respiration are of relatively greater importance than anthropogenic inputs in mobilizing metals. Chelation may also be significant where the effects of acid rain are not predominant (see references cited by Gorham et al. 1979).

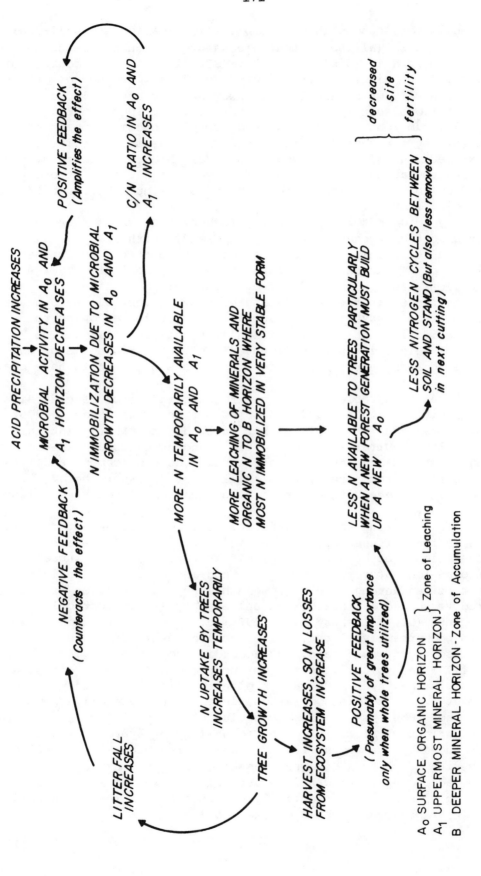

FIGURE 8.18 A hypothetical model of the effect of acid precipitation upon the nitrogen cycle in forests. SOURCE: After Tamm (1976).

The soils most susceptible to rapid acidification are well-drained brown forest soils (alfisols) that are sandy and noncalcareous but not already strongly acid (Wiklander 1973, 1974, 1979). Such coarse soils are moderately to highly saturated by "basic" cations such as Ca^{++}, Mg^{++}, K^+, and Na^+, which are leached away as they are replaced by "acid" cations such as hydrogen ions, aluminum ions, and hydroxy-aluminum ions. As the exchange complex becomes dominated increasingly by the "acid" cations, the pH of the soil declines. Fine-textured mineral soils, because of their high clay content, have a much greater cation-exchange capacity, which is usually strongly saturated by "basic" cations, and such soils are much better buffered against acidification.

Strongly acid podzol (spodosol) soil horizons--whether organic, with a high cation-exchange capacity, or sandy, with low cation-exchange capacity--are only slightly susceptible to further acidification. Their exchange sites are already dominated by the "acid" cations, and thus added hydrogen ions are more likely to pass through the system in the percolating waters. Nevertheless, because such soils are already impoverished, even slight losses owing to increased acidification may be critical for soil fertility. The influence of acid rain upon the impoverished lateritic soils of the tropics could also be very significant but does not seem to have been examined.

An interesting problem is the synergistic influence of neutral salts upon cation exchange under acid conditions. Wiklander (1975, 1979, cf. Abrahamsen et al. 1979) has made the important point that adding divalent neutral sulfates--often abundant in acid rain--to acid leaching solutions may retard significantly the replacement and loss of basic cations from already acid soils and thus favor the acidification of receiving waters. Monovalent chlorides have a lesser effect.

Addition of strong acids to precipitation will cause not only ion exchange on the surfaces of soil particles but also alteration of the particles themselves by weathering, either directly or by increasing the hydrogen-ion saturation of organic and inorganic soil colloids, which act as weathering agents (Loughnan 1969). Carbonates are weathered very readily indeed, but unless they are present only in small amounts (cf. Salisbury 1922, 1925) effects upon the soil are likely to be extremely slight. Aluminosilicate minerals are more slowly dissolved by acid rain, but according to Norton (1976, cf. Johnson 1979) the solubility of aluminum compounds such as gibbsite, amorphous aluminum hydroxide, and kaolinite increases rapidly below a pH of about 5.5. In this connection, lake waters generally show an order-of-magnitude rise in dissolved aluminum as pH falls from 5.5 to around 4 (Wright et al. 1976). The solubility of hydrous oxides of iron, produced by the weathering of iron-bearing minerals, is affected by a lowering of pH to a somewhat lesser degree than oxides of aluminum, according to Black (1967, see also Birkeland 1974). The dissolution of silica is essentially unaffected over the pH range 2 to 8 but rises rapidly above pH 8 (Birkeland 1974).

Increased leaching of potassium, calcium, aluminum, and magnesium due to ion exchange reactions with hydrogen ions is one of the most commonly observed effects of acid precipitation (Figure 8.19; also see review by Abrahamsen 1980). These increased rates of leaching appear to outstrip compensatory increases in weathering, reducing the exchangeable pool of the above cations. Sulfate is generally adsorbed in soil, largely in the B horizon (Farrell et al. 1980, Singh 1980, Figure 8.20).

Whitby and Hutchinson (1974) found that soils acidified to pH values below 4.0 by smelter fumigations in the Sudbury, Ontario area released sufficient aluminum into the soil solution to severely inhibit the establishment and elongation of seedling roots. Ulrich et al. (1980) have suggested that elevated aluminum concentrations caused the crown dieback in beech (Fagus sylvatica) and the failure of seedlings in the Solling project research forest in Germany. The direct role of aluminum in this response as opposed to the effect of severe droughts which occurred during the same period has not been established.

It is not clear at present whether the effects of acid precipitation on soil will result in long-term degradation of terrestrial ecosystems. At least one controlled 5-year study has revealed a significant depletion of cation exchange capacity in soils subjected to loading of 13.3 k. eq. H+ per ha of artificial acid rain (Farrell et al. 1980). Troedsson (1980) found declining quantities of Ca^{++}, Mg^{++}, and K^+ in Swedish soils over a 10-year period. Both theoretical and experimental studies suggest that such declines may be widespread in vulnerable soils after several decades to a few centuries of acid precipitation (Odén 1968, Reuss 1975b, Malmer 1976, Norton 1976, Tamm 1977, McFee 1978, Abrahamsen and Stuanes 1980).

Differences in effects of different levels of acidification upon trace metals in soils are even less well understood. The only detailed study is that of Tyler (1978), in which purely organic mor humus layers (Romell 1932, 1935, Lutz and Chandler 1946) from Swedish spruce forests were leached in the laboratory by simulated rain acidified to 5 different pH levels. Two sets of humus layers were examined, one far from and the other near to a brass mill. The latter exhibited very strong contamination by copper (695 x the control set, far from the pollution source) and zinc (124 x), with lesser enrichment of cadmium (24 x), lead (7.6 x), chromium (3.7 x), manganese (3.1 x), nickel (2.2 x), and vanadium (1.3 x). Simulated rains at pH values of 4.2, 3.4, 3.2, 3.0, and 2.8 were applied over 125 days in amounts totaling 625 ml/g of humus.

The simulated rain at pH 4.2, a value often reached in Swedish precipitation, released (over 125 days) substantial percentages of several trace metals in the control humus layers, notably manganese (44%), nickel (44%), cadmium (33%), and zinc (25%). Percentage release was much less for the other metals, copper (12%), vanadium (11%), chromium (8.9%), and lead (2.2%). With the notable exception of vanadium release (40%), percentage releases were generally less in the humus layers of the polluted site, as follows: manganese (4.1%), nickel (11%), cadmium (9.3%), zinc (18%), copper (1.7%), chromium

174

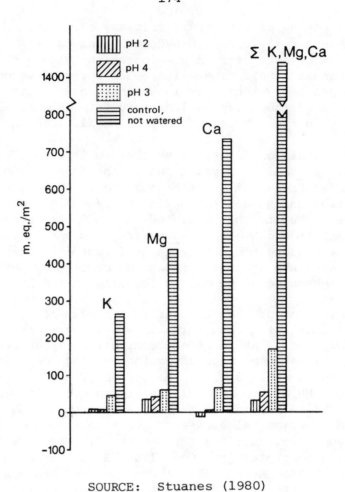

SOURCE: Stuanes (1980)

FIGURE 8.19 Loss of nutrients by weathering and mineralization. Artificial rain of varying acidity was applied in a field experiment over a 5-year period. The "rain" amounted to 50 mm per month for the 5 frost-free months of the year. SOURCE: Stuanes (1980).

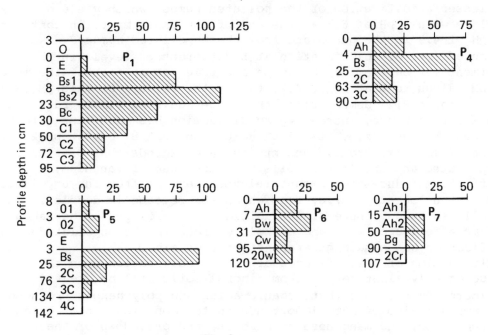

$^{35}SO_4^{2-}$ S adsorbed

FIGURE 8.20 Distribution of adsorbed $^{35}SO_4^{2-}$ sulfur in soil profiles. P_1 and P_5 are iron-podzols; P_4 is a semipodzol (all typic udipsamments); and P_6 and P_7 are brown-earths (umbric dystrochrept and aquic haploboroll, respectively). SOURCE: Singh (1980).

(1.5%), and lead (0.24%). This difference is most likely due to the much lesser acidification of the polluted humus, which yielded a percolate with a pH of 6.1 at the end of the experiment, in contrast to a pH of 4.5 in the percolate from the unpolluted humus.

The effects of acidification at 5 different pH levels in this experiment may be examined by comparing releases at each of the more acid pH values to that at pH 4.2, taking the release at this pH as unity. Figure 8.21 demonstrates that once again no simple rule can be given. For instance, increasing acidification has much less effect upon lead than upon zinc at all pH values above 2.8, but at pH 2.8 the release of lead is strongly accentuated and exceeds that of zinc in both polluted and unpolluted soils. In the case of vanadium, acidification reduces the amount released, except from the unpolluted humus at the lowest pH, 2.8, from which the release is doubled. In the polluted humus, increasing acidification below pH 4.2 has about the same effect upon zinc and nickel, whereas in the unpolluted humus, acidification has a much greater effect upon zinc.

Elucidation of the influence of acid precipitation upon the release of polyvalent metals from mineral soils will be greatly complicated by the fact that organic acids and polyphenols produced by organisms are also of much importance in the mobilization of oxides of aluminum, iron, and manganese from soils, and breakdown of these oxides will release the many trace metals adsorbed by them (several references in Russell 1973). Presumably acid rain--with its strong mineral acids--will have some effect upon the weathering action of the organic acids and polyphenols, the latter being known to reduce iron more strongly in acid than in neutral conditions (Russell 1973). Acidification can also reduce the stability of the fulvic acid components of humic acids in soils and their metal complexes, and thus metal availability should be greater at lower pH (Schnitzer 1980). Little is known about such interactions, however, and they should be given greater attention.

A number of soil microbial processes appear to be affected by acidification. Reduced soil pH may cause a reduction in nitrogen fixation (Alexander 1980), although mineralization of organic nitrogen may increase (Nyborg and Hoyt 1978). Increased acidity also causes nitrification to decrease; nitrification usually ceases entirely at about pH 4.0. There may be some degree of acclimation in acid environments (Walker and Wickramasinghe 1979). Denitrification, the main means by which nitrogen is released to the atmosphere from the biosphere, is also affected by acidification (Figure 8.22). Decreased rates, and a change in end product from molecular nitrogen to nitrous oxide, are observed in anoxic environments at pH values less than 6.0. At lower pH levels an appreciable yield of nitric oxide is observed, which is phytotoxic (Wijler and Delwiche 1954).

Only the mechanisms involved in nitrogen fixation have been studied directly. Legumes, which depend on Rhizobium for nitrogen fixation, are particularly sensitive to acidification. Alexander (1980) suggests that this may be due to the high concentrations of Al, Mn, or Fe in acid soils. Another possibility is limitation by

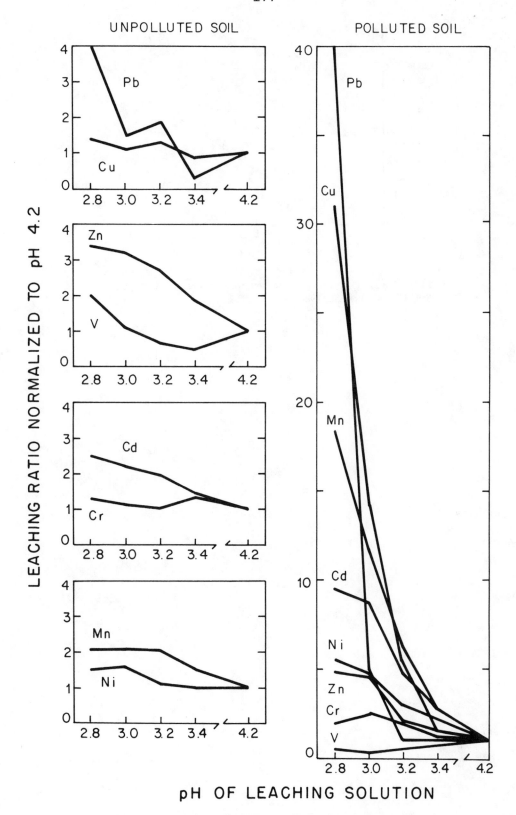

FIGURE 8.21 Effect of acidity upon the leaching of heavy metals from spruce humus layers, normalized to pH 4.2. Left-hand graphs represent unpolluted humus layers; right-hand graph represents humus layers polluted from a brass mill.

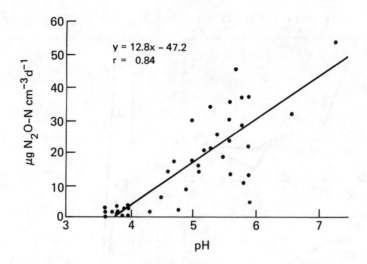

FIGURE 8.22 Correlation between denitrification rates and soil pH.
SOURCE: Müller et al. (1980).

molybdenum, an essential element for nitrogen fixation in legumes, because Mo solubility decreases at low pH.

Francis et al. (1980) found that increased acidity decreased rates of decomposition, ammonification, nitrification, denitrification, and N_2 fixation in soil. Microbial degradation of pesticides was also reduced at low pH.

Lohm (1980) found that the biomass of fungal mycelium increased in artifically acidified forest soils, replacing bacteria, which were reduced in number under acid conditions. Springtails also increased. The number of Enchytraeidae (oligochaete worms), Collembola (springtails), and Acari (mites) also decreased under acid conditions. The overall result was a net decrease in decomposition. Haagvar's (1980) study of invertebrates under acid conditions produced analogous results.

It appears that the detrimental effects of acidification on soils may not be due to--or at least, wholly attributable to--acidification per se but may be caused by the altered concentrations of nutrients and heavy metals in acidified soils. Direct toxicity of hydrogen ions in soils was found to be negligible over the pH range 4 to 8 (Arnon and Johnson 1942, Arnon et al. 1942); around pH 3 roots may be injured and above pH 8 phosphate absorption may be inhibited. In general, pH affects plant growth--and thereby the cycles of many (especially biophile) elements--by influencing the concentrations of different ions in the soil solution (Russell 1973, Nyborg 1978, Hutchinson and Collins 1978). Some of the elements affected are major plant nutrients (e.g., calcium, potassium, phosphate, nitrate), and others are toxicants (e.g., aluminum, lead), while several trace elements (e.g., manganese, copper, zinc) may be nutrients at low concentrations and toxicants at high concentrations (Bowen 1966), as for example around metal smelters (Stokes et al. 1973, Hutchinson and Whitby 1977).

The biogeochemistry of an ecosystem varies systematically as both vegetation and soil change in the course of succession, and acidification (whether normal or anthropogenic) plays a marked role in this process (Gorham et al. 1979). For a time, the leaching effect of acid precipitation upon a soil with appreciable reserves of calcium may enrich the stream and lake waters draining such soils without acidifying them (Gordon and Gorham 1963, Gorham 1978a), because the acids will be neutralized in percolating through the upland soils and liberating basic cations. However, if water flow occurs mainly as surface runoff, for example over frozen ground in spring, or downward through old root channels, animal burrows, and along rock faces (Tamm and Troedsson 1957)--then acidification of the receiving water may take place even if the upland soil possesses quite substantial buffering capacity. As time goes on and upland soils become more acid, they also become less susceptible to further acidification by acid rain, because exchange sites within the soil are already highly saturated by hydrogen ions. In this situation, substantial amounts of acid will percolate to the receiving waters of streams and lakes, which consequently undergo a pronounced decline in pH.

Acidification may be regarded as a normal tendency of ecosystem succession on base-poor substrata, because the biota produce acids of various kinds metabolically (carbonic acid and various organic acids through the oxidation of organic carbon compounds; nitric acid by nitrifying bacteria; sulfuric acid by bacterial oxidation of organic sulfur compounds; and hydrogen ions attached to the surfaces of roots, Sphagnum mosses, fungal hyphae, and bacteria, cf. Wiklander 1979). But natural acidification is accelerated appreciably by acid rain, which often brings about a change within decades in vulnerable aquatic ecosystems, once their buffering capacity is exhausted. Acid rain may also cause a slower change--perhaps over centuries--in upland ecosystems, where soil minerals and ion-exchange complexes provide a greater degree of buffering, even in the most vulnerable watersheds. If acidification should lead to substantial reduction in soil base saturation, recovery following removal of the acid loading could well take decades to centuries. Both natural and anthropogenic acidification processes in terrestrial soils deserve increased study, so that we may assess their relative importance in different ecosystems.

Wetlands

The disappearance of several species of the bog moss Sphagnum from the vast blanket bogs of the southern Pennines in the British Isles is the major vegetational change that has resulted from atmospheric pollution in that country since the Industrial Revolution (Tallis 1964). In this connection, Gorham (1958b) has shown that pools in bogs with an intact Sphagnum cover ranged in pH from 4.5 in remote areas of Britain, where the acidity is chiefly biogenic, to as low as 3.9 in the northern Pennines closer to urban/industrial centers. In these bog waters the correlation between H^+ and non-marine $\overline{\overline{SO_4}}$ ions was highly significant ($\underline{r}$ = 0.985). Near Sheffield in the southern Pennines the pH of bog pools was only 3.25, and the concentration of $\overline{\overline{SO_4}}$ reached 46 mg/l. Although it is impossible to ascribe with certainty the disappearance of the bog moss from the southern Pennines to acid precipitation, recent experiments with the same Sphagnum species indicate that either acid rain or SO_2 fumigation alone have detrimental effects consistent with the observed disappearance (Ferguson, Lee, and Bell 1978). In contrast, Hemond (1980) has calculated that in a Massachusetts peatland dominated by Sphagnum the reduction of sulfate and biological utilization of nitrate totally buffer the effect of acid deposition upon interstitial waters there.

Vast areas of sphagnum bog occur in North America, including portions of northern Minnesota and the lowlands of Hudson and James bays where rain is now quite acidic. Many of these peatland ecosystems have very low buffering capacities, which suggests that they may be easily unbalanced by acid precipitation. Although data are sparse, the abundance of peatlands in northern latitudes around the globe suggests that they may be significant reservoirs in the global cycles of many elements.

ASSESSMENT OF ECOLOGICAL EFFECTS

For improvements in our understanding of the effects of acid deposition we must concentrate on two major research areas. First, we need long-term monitoring (Botkin 1978) of sensitive organisms or communities (e.g., the "neuston" of freshwater surfaces, cf. Gorham 1976, 1978a,b) in especially vulnerable ecosystems, so that we can receive early warning of ecosystems at hazard. Such monitoring presupposes the identification of a series of indicator organisms and communities (Thomas 1972) or indices to community structure (Cairns 1974) and the development of an adequate scheme for rating ecosystem vulnerability. The Calcite Saturation Index developed by Conroy et al. (1974, see also Kramer 1976) provides a useful guide to the vulnerability of lakes, but because it uses only the concentrations of calcium, bicarbonate, and hydrogen ions in the water, it does not take into account the full range of factors involved (cf. Table 8.1).

Second, we need experimental ecosystem-scale studies to examine linkages among and processes within uplands, wetlands, streams, and lakes (Likens and Bormann 1974, Schindler 1980b) and the mechanisms by which acid deposition alters ecosystem function. Some of these studies could be conducted on watersheds exposed to ambient levels of acid rain, while other studies, in relatively unpolluted areas, could subject whole ecosystems to experimental acidification. Perhaps the recently proposed national network of experimental ecological reserves (TIE 1977) could provide suitable sites for such studies, as could several of the experimental watersheds (Table 22 in Likens et al. 1977) set up for other purposes. One of these, the Hubbard Brook Experimental Forest, has already become an important site for research on acid rain (Likens et al. 1977).

AMELIORATION

Of the options presently available only the control of emissions of sulfur and nitrogen oxides can significantly reduce the rate of deterioration of sensitive freshwater ecosystems. It is desirable to have precipitation with pH values no lower than 4.6 to 4.7 throughout such areas, the value at which rates of degradation are detectable by current survey methods, as mentioned above. In the most seriously affected areas (average precipitation pH of 4.1 to 4.2), this would mean a reduction of 50 percent in deposited hydrogen ions. Control of SO_2 from new electrical generating plants alone would be insufficient to accomplish this, and thus restrictions on older plants must be considered. Furthermore, there are no proposed restrictions on the emission of nitrogen oxides, and the amounts of these substances emitted are expected to continue to increase (see Figure 4.6).

The alkalinity of waters endangered by acidification can be enhanced by a number of means, most notably by "liming"--adding calcium carbonate or oxide--and by adding phosphorus to stimulate biological fixation of nitrate and CO_2. All of these techniques are expensive ($50 and more per hectare of water surface), and treatments must be repeated every few years. Due to high costs and logistic difficulties, lime cannot be applied to the vast areas that are currently endangered by acidification. In the areas most susceptible to acid deposition, it will therefore be impossible to maintain the alkalinity and pH of more than a few selected bodies of water. Furthermore, pH and alkalinity cannot be artificially maintained without increasing the ionic concentration of the receiving water, the consequences of which have not been investigated. The fate of dissolved toxic metals after liming is also poorly known.

Addition of nutrients to increase alkalinity has been investigated at a few sites in Ontario, both by the Ontario Ministry of Environment and Canadian Department of Fisheries and Oceans, but results are not available yet. Many of the objections to liming will also apply to nutrient additions.

SUMMARY

Acid deposition, due to the further oxidation of sulfur and nitrogen oxides released to the atmosphere by anthropogenic sources, is causing widespread damage to aquatic ecosystems, including loss of bicarbonate, increased acidity, and higher concentrations of toxic metals. As a result, several important species of fish and invertebrates have been eliminated over substantial parts of their natural ranges.

Effects on terrestrial ecosystems are less pronounced. Increased leaching of both nutrients and toxic elements is evident in poorly buffered soils sensitive to acidification. There is some evidence for damage to crop plants, and many soil microbial processes are negatively affected at low pH. Trees appear to be slightly stimulated by acid precipitation, although this effect is expected to be short-lived, because of increased leaching of cationic nutrients and the buildup of toxic concentrations of metals in soil water.

Better long-term studies of deposition processes and of effects on ecosystems are required to illuminate the complex ecological effects of acid precipitation and associated nutrients and toxicants. The control of emissions of sulfur and nitrogen oxides from fossil fuels is necessary to halt the acidification of sensitive aquatic ecosystems.

APPENDIX

BIOGRAPHICAL SKETCHES OF COMMITTEE MEMBERS

DAVID W. SCHINDLER is director of the Experimental Limnology Project of the Freshwater Institute, Canadian Department of Fisheries and Oceans in Winnipeg, Manitoba. He received his D. Phil. in ecology in 1966 from Oxford University, where he was a Rhodes Scholar. He is also an adjunct professor of Zoology at the University of Manitoba and Visiting Senior Research Associate at Lamont-Doherty Geological Observatory, Palisades, New York. Dr. Schindler's research specialty is limnology.

MARTIN ALEXANDER is Liberty Hyde Bailey Professor of Soil Science at Cornell University. He received his doctorate in bacteriology in 1955 from the University of Wisconsin and then joined the faculty at Cornell. Dr. Alexander specializes in nitrogen transformations in soils and water and other aspects of biochemical ecology and soil microbiology.

EDWARD D. GOLDBERG is Professor of Chemistry at Scripps Institution for Oceanography, University of California, San Diego. He joined the faculty at Scripps following the completion of his doctoral studies in chemistry at the University of Chicago in 1949. His research interests are in the geochemistry of marine waters and sediments and in atmospheric and marine pollution. Dr. Goldberg is a member of the National Academy of Sciences.

EVILLE GORHAM is Professor of Ecology and Botany at the University of Minnesota. He received his doctoral degree in plant ecology from the University of London, where he was an 1851 Exhibitioner and Keddey Fletcher-Warr Scholar. Afterward, he held a Royal Society of Canada Research Fellowship at the State Forest Research Institute in Stockholm. He has taught at the universities of London, Toronto, and Calgary and has served on the staff of the Freshwater Biological Association in the English Lake District. Dr. Gorham's research interests are in wetland ecology, limnology, and biogeochemistry, with particular emphasis on atmospheric deposition.

DANIEL GROSJEAN is manager of the Environmental Chemistry Center, Environmental Research and Technology, Inc. (ERT), Westlake Village, California, and holds an appointment as Visiting Associate in the Division of Chemical Engineering, California Institute of Technology. His doctoral degree in physical organic chemistry is from the University of Paris. Dr. Grosjean's research programs are concerned with the characterization and chemical transformations of gaseous and particulate pollutants in the atmosphere with emphasis on organic toxic substances.

HALSTEAD HARRISON is an Associate Professor in the Atmospheric Sciences Department of the University of Washington. He received his Ph.D. in chemistry from Stanford University in 1960, and he has held an NSF fellowship at the Institute for Applied Physics in Bonn, West Germany, and research positions with the General Atomic Division of General Dynamics Corporation and the Boeing Science Research Laboratories. Dr. Harrison's area of research specialization is atmospheric chemistry and applied mathematics.

WALTER W. HECK is Professor of Botany at North Carolina State University and research leader for the Air Quality Research Program, USDA-SEA/ Agriculture Research. He received his doctoral degree in plant physiology in 1954 from the University of Illinois and taught at Texas A & M University before joining the USDA Agricultural Research Service. Dr. Heck studies the effects of environmental stress on the response of plants to air pollutants. He has had an active research program in the area of air pollution effects on vegetation for twenty-two years.

RUDOLPH B. HUSAR is Professor of Mechanical Engineering at Washington University. Following the completion of his doctoral degree work in mechanical engineering at the University of Wisconsin in 1971 he spent two years as a visiting professor at the Meteorological Institute of the University of Stockholm. Dr. Husar's research is concerned with the modeling of atmospheric pollutants.

THOMAS C. HUTCHINSON is Professor of Botany and Forestry and Associate of the Institute for Environmental Studies at the University of Toronto. He received his doctoral degree from the University of Sheffield in 1966. Dr. Hutchinson is chairman of the Heavy Metals Panel of the Canadian Research Council. His research interests include the environmental effects and phytotoxicity of heavy metals, air and water pollution (with particular interest in acid precipitation), and the impacts of pollutants in arctic ecosystems.

SVANTE ODEN is Professor of Soil Science and Ecochemistry at the Swedish University of Agricultural Sciences, Uppsala. Dr. Odén is known for his work on concepts of acid precipitation and its effects on aquatic and terrestrial ecosystems.

GERALD T. ORLOB is Professor of Civil Engineeering, University of
California at Davis. He received his doctoral degree in hydraulic
engineering from Stanford University in 1959. Dr. Orlob's research
interests are in the field of water resource management, particularly
directed to the formulation and application of mathematical models for
simulation of hydromechanical, and ecological behavior of streams,
lakes, estuaries, and coastal waters.

LARS OVERREIN, former director of the Norwegian National Research Program
on Acid Precipitation (the SNSF Project), is currently Director General
of the Norwegian Water Research Institute, Oslo, Norway.

DOUGLAS M. WHELPDALE heads the program on long-range transport of
atmospheric pollutants for the Atmospheric Environment Service,
Environment Canada. He received his doctoral degree in atmospheric
physics from the University of Toronto in 1970, and his research is on
the long-range transport of air pollutants and their removal from the
atmosphere. He currently serves as a member of the Canadian-U.S.
research group studying problems of long-range transport and as a
member of the steering committee of the European Monitoring and
Evaluating Program which studies similar problems.

REFERENCES

Abeles, F.B., L.E. Craker, and L.E. Forrence (1971) Fate of air pollutants: Removal of ethylene, sulfur dioxide, and nitrogen dioxide by soil. Science 173:914-916.

Abrahamsen, G., and G.J. Dollard (1979) Effects of acid precipitation on forest vegetation and soil. Section 4.2 (17 pages), Ecological Effects of Acid Precipitation, EPRI Report No. EA-79-6-LD. Palo Alto, CA: Electric Power Research Institute.

Abrahamsen, G., A.O. Stuanes, and K. Bjor (1979) Interaction between simulated rain and barren rock surface. Water, Air, Soil Pollut. 11:191-200.

Abrahamsen, G. (1980) Acid precipitation, plant nutrients and forest growth. Pages 58-63, Ecological Impact of Acid Precipitation. Proceedings of an International Conference, Sandefjord, Norway, March 11-14, 1980, edited by D. Drablös and A. Tollan. Oslo-Aas, Norway: SNSF project.

Abrahamsen, G., and A.O. Stuanes (1980) Effects of simulated rain on the effluent from lysimeters with acid, shallow soil, rich in organic matter. Pages 152-153, Ecological Impact of Acid Precipitation. Proceedings of an International Conference, Sandefjord, Norway, March 11-14, 1980, edited by D. Drablös and A. Tollan. Oslo-Aas, Norway: SNSF project.

Adams, D.F., and S.O. Farwell (1980) Biogenic sulfur gas emissions from soils in eastern and southeastern United States. Paper 80-40.5, 73rd Annual Meeting of Air Pollut. Control Association, Montreal, Quebec, June 1980. Pittsburgh, PA: Air Pollution Control Association.

Adams, J.A.S., M.S.M. Mantovani, and L.L. Lundell (1977) Wood versus fossil fuel as a source of excess carbon dioxide in the atmosphere: A preliminary report. Science 196:54-56.

Aiton, W. (1811) Treatise on the Origin, Qualities, and Cultivation of Moss-Earth, with Directions for Converting it into Manure. Air, Scotland: Wilson and Paul.

Alexander, M. (1980) Effects of acid precipitation on biochemical activities in soil. Pages 47-52, Ecological Impact of Acid Precipitation. Proceedings of an International Conference,

Sandefjord, Norway, March 11-14, 1980, edited by D. Drablös and A. Tollan. Oslo-Aas, Norway: SNSF project.

Alexander, M. (1981) Biodegradation of chemicals of environmental concern. Science 211:132-138.

Allen, H.E., R.H. Hall, and T.D. Brisbin (1980) Metal speciation: Effects on aquatic toxicity. Environ. Sci. Technol. 14:441-442.

Allen, W.R., and P.M. Sheppard (1971) Copper tolerance in some Californian populations of the Monkey flower Mimulus guttatus. Proc. R. Soc. Lond. [Biol.] 177:177-196.

Allsup, J.R., and D.B. Eccleston (1980) Ethanol/Gasoline Blends as Automobile Fuels. Bartlesville, OK: U.S. Department of Energy, Bartlesville Energy Technology Center.

Almer, B., W. Dickson, C. Ekström, E. Hörnström, and U. Miller (1974) Effects of acidification on Swedish lakes. AMBIO 3:30-36.

Almer, B., W. Dickson, C. Ekström, and E. Hörnström (1978) Sulfur pollution and the aquatic ecosystem. Pages 271-311, Sulfur in the Environment, Part II, edited by J.O. Nriagu. New York: John Wiley & Sons.

Andreae, M.O. (1980) Dimethylsulfoxide in marine and fresh waters. Limnol. and Oceanography 25:1054-1063.

Andren, A.W., A.W. Elzerman, and D.E. Armstrong (1976) Chemical and physical aspects of surface organic microlayers in freshwater lakes. J. Great Lakes Res. 2(Suppl. 1):101-110.

Aneja, V.P., J.T. Overton, Jr., and A.P. Aneja (1980) The effect of moisture on the release of biogenic sulfur compounds. Paper 80-40.4, 73rd Annual Meeting of Air Pollut. Control Association, Montreal, Quebec, June 1980. Pittsburgh, PA: Air Pollution Control Association.

Armstrong, F.A.J. (1979) Mercury in the aquatic environment. Pages 84-100, Effects of Mercury in the Canadian Environment. Publication NRCC No. 16739. Ottawa: National Research Council Canada.

Arnon, D.I., and C.M. Johnson (1942) Influence of hydrogen ion concentration on the growth of higher plants under controlled conditions. Plant Physiol. 17:525-539.

Arnon, D.I., W.E. Fratzke, and C.M. Johnson (1942) Hydrogen ion concentration in relation to absorption of inorganic nutrients by higher plants. Plant Physiol. 17:515-524.

Atkins, W.R.G. (1947) The electrical conductivity of river, rain and snow water. Nature 159:674.

Atkinson, R., K.R. Darnall, A.C. Lloyd, A.M. Winer, and J.N. Pitts, Jr. (1979) Kinetics and mechanisms of the reactions of the hydroxyl radical with organic compounds in the gas phase. Pages 375-487, Advances in Photochemistry, Vol. 11, edited by J.N. Pitts, G.S. Hammond, K. Gollnick, and D. Grosjean. New York: John Wiley and Sons.

Atkinson, R., W.P.L. Carter, K.R. Darnall, A.M. Winer, and J.N. Pitts, Jr. (1980) A smog chamber and modeling study of the gas phase NO-air photooxidation of toluene and the cresols. Int. J. Chem. Kinet. 12:779-836.

Atlas, E., and C.S. Giam (1981) Global transport of organic
    pollutants: Ambient concentrations in the remote marine
    atmosphere. Science 211:163-165.

Aulie, R.P. (1973) Jean Baptiste Joseph Deudonne Boussingault. Dict.
    Sci. Biogr. 2:356-357.

Avenhaus, R. (1977) Material Accountability: Theory, Verification,
    and Applications. New York: Wiley-Interscience.

Ayres, R.U. (1978) Resources, Environment, and Economics. New York:
    Wiley-Interscience.

Bache, B.W., and J.E. Rippon (1979) Study group: 3 discussions.
    Section 4.3 (5 pages), Ecological Effects of Acid Precipitation,
    EPRI Report No. EA-79-6-LD. Palo Alto, CA: Electric Power
    Research Institute.

Bailey, L.D., and E.G. Beauchamp (1973) Effects of temperature on
    nitrate and nitrite ion reduction, nitrogenous gas production, and
    redox potential in a saturated soil. Can. J. Soil Sci. 53:213-218.

Baker, J.J.W., and G.E. Allen (1971) The Study of Biology. 2nd ed.
    Reading, MA: Addison-Wesley.

Baker, J.P., and C.L. Schofield (1980) Aluminum toxicity to fish as
    related to acid precipitation and Adirondack surface water
    quality. Pages 292-293, Ecological Impact of Acid Precipitation.
    Proceedings of an International Conference, Sandefjord, Norway,
    March 11-14, 1980, edited by D. Drablös and A. Tollan. Oslo-Aas,
    Norway: SNSF project.

Banwart, W.L., and J.M. Bremner (1975) Identification of sulfur gases
    evolved from animal manures. J. Environ. Qual. 4:363-366.

Banwart, W.L., and J.M. Bremner (1976a) Evolution of volatile sulfur
    compounds from soils treated with sulfur-containing organic
    materials. Soil Biol. Biochem. 8:439-443.

Banwart, W.L., and J.M. Bremner (1976b) Volatilization of sulfur from
    unamended and sulfate-treated soils. Soil Biol. Biochem. 8:19-22.

Barkman, J.J. (1958) Phytosociology and Ecology of Cryptogamic
    Epiphytes. Assen, Netherlands: Van Gorcum.

Barrett, E., and G. Brodin (1955) The acidity of Scandinavian
    precipitation. Tellus 7:251-257.

Barth, D.S., and S.M. Blacker (1978) The EPA program to assess the
    public health significance of diesel emissions. J. Air Pollut.
    Control Assoc. 28:769-771.

Bartholomew, G.W., and M. Alexander (1979) Microbial metabolism of
    carbon monoxide in culture and in soil. Appl. Environ.
    Microbiol. 37:932-937.

Beauford, W., J. Barber, and A.R. Barringer (1975) Heavy metal release
    from plants into the atmosphere. Nature 256:35-37.

Beauford, W., J. Barber, and A.R. Barringer (1977) Release of
    particles containing metals from vegetation into the atmosphere.
    Science 195:571-573.

Beijer, K., and A. Jernelöv (1978) Ecological aspects of
    mercury-selenium interactions in the marine environment. Environ.
    Health Perspect. 25:43-45.

Bennett, J.H., A.C. Hill, and D.M. Gates (1973) A model for gaseous
    pollutant uptake by plants. J. Air Pollut. Control Assoc.
    23:957-962.

Benson, A.A., and R.E. Summons (1981) Arsenic accumulation in Great Barrier Reef invertebrates. Science 211:482-483.

Berger, W.H., and G.R. Heath (1968) Vertical mixing in pelagic sediments. J. Marine Res. 268:134-143.

Bertine, K.K., and E.D. Goldberg (1971) Fossil fuel combustion and the major sedimentary cycle. Science 173:233-235.

Bertine, K.K., and E.D. Goldberg (1977) History of heavy metal pollution in Southern California coastal zone-reprise. Environ. Sci. Technol. 11:297-299.

Birkeland, P.W. (1974) Pedology, Weathering, and Geomorphological Research. New York: Oxford University Press.

Biswas, A.K. (1970a) Edmond Halley, F.R.S., hydrobiologist extraordinary. Notes and Records, R. Soc. Lond. 25:47-57.

Biswas, A.K. (1970b) History of Hydrology. Amsterdam: North Holland Publishing Company.

Black, A.B. (1967) Applications: Electrokinetic characteristics of hydrous oxides of aluminum and iron. Pages 247-300, Principles and Applications of Water Chemistry, edited by S.D. Faust and J.V. Hunter. New York: John Wiley & Sons.

Blackmer, A.M., and J.M. Bremner (1976) Potential of soil as a sink for atmospheric nitrous oxide. Geophys. Res. Lett. 3:739-742.

Block, C., and R. Dams (1976) Study of fly ash emissions during combustion of coal. Environ. Sci. Technol. 10:1011-1017.

Bloomfield, C. (1969) Sulfate reduction in waterlogged soils. J. Soil Sci. 20:207-221.

Blumer, M., and W.W. Youngblood (1975) Polycyclic aromatic hydrocarbons in soils and recent sediments. Science 188:53-55.

Blumer, M., W. Blumer, and T. Reich (1977) Polycyclic aromatic hydrocarbons in soils of a mountain valley: Correlated with highway traffic and cancer incidence. Environ. Sci. Technol. 11:1082-1084.

Boethling, R.S., and M. Alexander (1979a) Microbial degradation of organic compounds at trace levels. Environ. Sci. Technol. 13:989-991.

Boethling, R.S. and M. Alexander (1979b) Effect of concentration of organic chemicals on their biodegradation by natural microbial communities. Appl. Environ. Microbiol. 37:1211-1238.

Bolin, B., ed. (1971) Sweden's National Report to the United Nations Conference on the Human Environment: Air pollution Across National Boundaries. The Impact on the Environment of Sulfur in Air and Precipitation. Stockholm: Norsted.

Bolin, B., and R.J. Charlson (1976) On the role of the tropospheric sulfur cycle in the shortwave radiative climate of the earth. AMBIO 5:47-54.

de Bont, J.A.M., K.K. Lee, and D.F. Bouldin (1978) Bacterial oxidation of methane in a rice paddy. Pages 91-96, Environmental Role of Nitrogen-fixing Blue-green Algae and Asymbiotic Bacteria. Ecological Bulletins. Stockholm: Swedish National Research Council.

Bossard, P., and Gächter, R. (1979) Effects of increased heavy metal lead on uptake of glucose by natural planctonic communities. Schweiz. Z. Hydrol. 41:261-270.

191

Botkin, D.B. (1977) Long-Term Ecological Measurement: Report of a
    Conference. Washington, D.C.: National Science Foundation.
Botkin, D.B., ed. (1978) A Pilot Program for Long-Term Observation
    and Study of Ecosystems in the United States: Report to the
    National Science Foundation (Program in Biological Research
    Resources). Washington, DC: National Science Foundation.
Bottini, O. (1939) Le pioggie caustiche nella regione vesuviana. Ann.
    Chim. Applic. 29:425-433.
Boutron, D., and R. Delmas (1980) Historical record of global
    atmospheric pollution revealed in polar ice sheets. AMBIO
    9:210-215.
Bowen, H.J.M. (1966) Trace Elements in Biochemistry. New York:
    Academic Press.
Bradbury, J.P., and J.C.B. Waddington (1973) The impact of European
    settlement on Shagawa Lake, northeastern Minnesota, USA. Pages
    289-307, Quaternary Plant Ecology, edited by H.J.B. Birks and R.G.
    West. Oxford: Blackwell.
Bradbury, J.P. (1975) Diatom Stratigraphy and Human Settlement in
    Minnesota. Geological Society of America Special Paper, No. 171.
    Denver, CO: Geological Society of America.
Bradshaw, A.D. (1952) Populations of Agrostis tenuis resistant to lead
    and zinc poisoning. Nature 169:1098.
Braman, R.S., and C.C. Foreback (1973) Methylated forms of arsenic in
    the environment. Science 182:1247.
Bremner, J.M., and W.L. Banwart (1976) Sorption of sulfur gasses by
    soils. Soil Biol. Biochem. 8:79-83.
Bremner, J.M., and A.M. Blackmer (1978) Nitrous oxide: Emissions from
    soils during nitrification of fertilizer nitrogen. Science
    199:295-296.
Brezonik, P.L. (1977) Denitrification in natural waters. Prog. Water
    Technol. 8:373-392.
Brimblecombe, P. (1975) Industrial air pollution in thirteenth-century
    Britain. Weather 30:388-396.
Brimblecombe, P. (1976) Attitudes and responses towards air pollution
    in medieval England. J. Air Pollut. Control Assoc. 26:941-945.
Brimblecombe, P. (1977) London air pollution, 1500-1900. Atmos.
    Environ. 11:1157-1162.
Brimblecombe, P., and C. Ogden (1977) Air pollution in art and
    literature. Weather 32:285-291.
Brimblecombe, P. (1978) Air pollution in industrializing England.
    J. Air Pollut. Control Assoc. 28:115-118.
Brimblecombe, P., and J. Pitman (1980) Long-term deposit at
    Rothamsted, Southern England. Tellus 32:261-267.
Brock, T.D. (1961) Milestones in Microbiology. Englewood Cliffs, NJ:
    Prentice Hall.
Broecker, W.S. (1974) Chemical Oceanography. New York: Harcourt,
    Brace, Jovanovich.
Broecker, W.S., and T.H. Peng (1974) Gas exchange rates between air
    and sea. Tellus 26:21-35.
Brown, D.A., and T.R. Parsons (1978) Relationship between cytoplasmic
    distribution of mercury and toxic effects to zooplankton and chum

salmon (Onchorkynchus keta) exposed to mercury in a controlled
ecosystem. J. Fish. Res. Bd. Can. 35:880-884.

Brown, D.A. (In press) Trace metals. The biological effects of stress
in estuarine environments: a manual, edited by B.L. Bayne. New
York: Holt, Rinehart and Winston/CBS.

Brown, G.W., Jr. (1976) Effects of polluting substances on enzymes of
aquatic organisms. J. Fish. Res. Bd. Can. 33:2018-2022.

Brown, H. (1975) Population growth and affluence: The fissioning of
human society. Q. J. Econ. 89:236-246.

Brown, H. (1976) Energy in our future. Pages 1-36, Annual Review of
Energy, Vol. 1, edited by J.M. Hollander and M.K. Simmons. Palo
Alto, CA: Annual Reviews.

Brunskill, G.J., B.W. Graham, and J.W.M. Rudd (1980) Experimental
studies of the effect of arsenic on microbial degradation of
organic matter and algal growth. Can. J. Fish Aquat. Sci.
37:415-423.

Bumb, R.R., W.B. Crummett, S.S. Cutie, J.R. Gledhill, R.H. Hummel,
R.O. Kagel, L.L. Lamparski, E.V. Luoma, D.L. Miller, T.J.
Nestrick, L.A. Shadoff, R.H. Stehl, and J.S. Woods (1980) Trace
chemistries of fire: A source of chlorinated dioxins. Science
210:385-390.

Burk, R.F., K.A. Foster, P.M. Greenfield, and K.W. Kiker (1974)
Binding of simultaneously administered inorganic selenium and
mercury to a rat plasma protein. Proc. for the Soc. for
Experimental Biol. and Medicine 145:782-785.

Butcher, S.S., and E.M. Sorenson (1979) A study of wood stove
particulate emissions. J. Air Pollut. Control Assoc. 29:724-728.

Butler, G.C., ed. (1978) Principles of Ecotoxicology. SCOPE Report
No. 12. New York: John Wiley and Sons.

Cairns, J., Jr. (1974) Indicator species versus the concept of
community structure as an index of water pollution. Water
Resources Bull. 10:338-347.

Calamari, D., and R. Marchetti (1973) The toxicity of mixtures of
metals and surfactants to rainbow trout (Salmo gairdneri Rich.)
Water Res. 7:1453-1464.

Caldecott, R.S., and L. Snyder, eds. (1960) Radioisotopes in the
Biosphere. Minneapolis: University of Minnesota Press.

Callendar, G.S. (1938) The artificial production of carbon dioxide and
its influence on temperature. Q. J. R. Meteorol. Soc. 64:223-237.

Callendar, G.S. (1949) Can carbon dioxide influence climate? Weather
4:310-314.

Carson, R. (1962) Silent Spring. Boston: Houghton Mifflin.

Carter, A. (1978) The Tolerance of Soil Fungi to Elevated Nickel
Levels in the Sudbury Area. M.Sc. Thesis, Department of Botany,
University of Toronto, Toronto.

Carter, W.P.L., A.C. Lloyd, J.L. Sprung, and J.N. Pitts, Jr. (1979a)
Computer modeling of smog chamber data: Progress in validation of
a detailed mechanism for the photooxidation of propene and
n-butane in photochemical smog. Int. J. Chem. Kinet. 11:45-101.

Carter, W.P.L., K.R. Darnell, R.A. Graham, A.M. Winer, and J.N.
Pitts, Jr. (1979b) Reactions of C_2 and C_4 a - Hydroxy
radicals with oxygen. J. Phys. Chem. 83:2305-2311.

Challenger, F. (1951) Biological methylation. Adv. Enzymol. 12:429-491.

Chamberlain, A.C. (1968) Transport of gases to and from surfaces with bluff and wave-like roughness elements. Quarterly Journal of the Royal Meterological Society. 94:318.

Chamberlain, A.C. (1975) The movement of particles in plant communities. Pages 155-201, Vegetation and the Atmosphere, Vol. 1, edited by Montieth. New York: Academic Press.

Chameides, W.L., D.H. Stedman, R.R. Dickerson, R.W. Rusch, and R.J. Cicerone (1977) NO_x production in lightning. J. Atmos. Sci. 34:143-149.

Chan, Y.K., and N.E.R. Campbell (1980) Denitrification in Lake 227 during summer stratification. Can. J. Fish. Aquat. Sci. 37:506-512.

Chatfield, R., and H. Harrison (1977) Tropospheric ozone. 2. Variations along a meridional band. J. Geophys. Res. 82:5969-5976.

Chau, Y.K., P.T.S. Wog, B.A. Silverberg, P.L. Luxon, G.A. Bengert (1976) Methylation of selenium in the aquatic environment. Science 192:1130.

Cheng, C.N., and D.D. Focht (1979) Production of arsine and methylarsines in soil and in culture. Appl. Environ. Microbiol. 38:494-498.

Cherian, M.G., and R.A. Goyer (1978) Metallothioneins and their role in the metabolism and toxicity of metals. Life Sciences 23:1-10.

Cleveland, W.S., and T.E. Graedel (1979) Photochemical air pollution in the northeast United States. Science 204:1273-1278.

Cloud, P.E., Jr. (1968) Atmospheric and hydrospheric evolution on the primitive earth. Science 160:729-736.

Cloud, P.E., Jr. (1976) Beginnings of biospheric evolution and their biogeochemical consequences. Paleobiology 2:351-387.

Clymo, R.S. (1964) The origin of acidity in Sphagnum bogs. The Bryologist 67:427-431.

Cogbill, C.V., and G.E. Likens (1974) Acid precipitation in the northeastern United States. Water Resources Res. 10:1133-1137.

Cohen, J.B., and A.G. Ruston (1912) Smoke. London: Arnold.

Collard, P. (1976) The Development of Microbiology. New Rochelle, NY: Cambridge University Press.

Congiu, L., F.T. Corongiu, M. Dore, C. Montaldo, S. Vargiolu, D. Casula, and G. Spiga (1979) The effect of lead nitrate on the tissue distribution of mercury in rats treated with methyl mercury chloride. Toxicol. and Appl. Pharm. 51:363-366.

Conroy, N., D.S. Jeffries, and J.R. Kramer (1974) Acid Shield Lakes in the Sudbury, Ontario Region. Pages 45-61, Proceedings, 9th Symposium on Water Pollution Research in Canada. Rexdale, Ontario: Ontario Ministry of the Environment.

Cook, R.B. (1981) The biogeochemistry of sulfur in two small lakes. Ph.D. Thesis. Department of Geology. New York: Columbia University.

Cooper, J.A. (1980) Environmental impact of residential wood combustion emissions and its implications. J. Air Pollut. Control Assoc. 30:855-861.

Costonis, A.C. (1970) Acute foliar injury of eastern white pine induced by sulfur dioxide and ozone. Phytopathology 60:994-999.

Costonis, A.C. (1972) Effects of ambient sulfur dioxide and ozone on eastern white pine in a rural environment. Phytopathology 61:717-720.

Costonis, A.C. (1973) Injury to eastern white pine by sulfur dioxide and ozone alone and in mixtures. Eur. J. Forest Pathol. 3:50-55.

Cottam, C., and E. Higgins (1946) DDT: Its Effect on Fish and Wildlife. Fish and Wildlife Service Circular 11. Washington, DC: U.S. Department of the Interior.

Cowling, E.B. (1981) An Historical Resumé of Progress in Scientific and Public Understanding of Acid Precipitation and Its Biological Consequences. SNSF Project Research Report Series. Oslo-Aas, Norway: SNSF project.

Cox, D.P., and M. Alexander (1973) Production of trimethylarsine gas from various arsenic compounds by three sewage fungi. Bull. Environ. Contam. Toxicol. 9:84-88.

Cox, R.M., and T.C. Hutchinson (1979) Metal co-tolerances in the grass Deschampsia cespitosa. Nature 279:231-233.

Cox, R.M., and T.C. Hutchinson (1980) Multiple metal tolerances in the grass Deschampsia cespitosa (L.) Beauv. from the Sudbury smelting area. New Phytologist 84:631-647.

Craig, G.R., and W.F. Baksi (1977) The effects of depressed pH on flagfish reproduction, growth and survival. Water Res. 11:621-626.

Crecelius, E.A., M.H. Bothner, and R. Carpenter (1975) Geochemistries of arsenic, antimony, mercury, and related elements in sediments of Puget Sound. Environ. Sci. Technol. 9:325-333.

Crisman, T.L., R.L. Schulze, P.L. Brezonik, and S.A. Bloom (1980) Acid precipitation: The biotic response in Florida lakes. Pages 296-297, Ecological Impact of Acid Precipitation. Proceedings of an International Conference, Sandefjord, Norway, March 11-14, 1980, edited by D. Drablös and A. Tollan. Oslo-Aas, Norway: SNSF project.

Cronan, C.S., and C.L. Schofield (1979) Aluminum leaching response to acid precipitation: Effects on high elevation watersheds in the Northeast. Science 204:304-306.

Crosland, M.P. (1973) Pierre Eugene Marcellin Berthelot. Dict. Sci. Biogr. 2:63-72.

Crowther, C., and A.G. Ruston (1911) The nature, distribution and effects upon vegetation of atmospheric impurities in and near an industrial town. J. Agric. Sci. 4:25-55.

Crowther, C., and D.W. Steuart (1913) The distribution of atmospheric impurities in the neighborhood of an industrial city. J. Agr. Sci. 5:391-408.

Crutzen, P.J. (1970) The influence of nitrogen oxides on the atmospheric ozone content. Q. J. R. Meteorol. Soc. 96:320-325.

Crutzen, P.J. (1979) The role of NO and NO_2 in the chemistry of the troposphere and stratosphere. Annu. Rev. Earth Planet. Sci. 7:443-472.

Crutzen, P.J., L.E. Heidt, J. Krasnec, W.H. Pollock, and W. Seiler (1979) Biomass burning as a source of the atmospheric gases CO, H_2, N_2O, NO, CH_3Cl, and COS. Nature 282:253-256.

195

Curtin, G.C., H.D. King, and E.L. Mosier (1974) Movement of elements into the atmosphere from coniferous trees in subalpine forests of Colorado and Idaho. J. Geochem. Explor. 3:245-263.

Curtis, C. ed. (1980) Before the rainbow: What we know about acid rain. Vol. 9. Washington, DC: Edison Electric Institute.

Czuba, M., and D.P. Ormood (1975) Effects of cadmium and zinc on ozone-induced phototoxicity in cress and lettuce. Can. J. Botany 52:645-649.

Dampier, W.C. (1948) A History of Science. 4th ed. New Rochelle, NY: Cambridge University Press.

Danckwerts, P.V. (1970) Gas-Liquid Reactions. New York: McGraw Hill Book Company.

Darwin, C. (1846) An account of the fine dust which often falls on vessels in the Atlantic Ocean. Q. J. Geol. Soc. Lond. 2:26-30.

Dau, H.C. (1823) Neues Handbuch über den Torf, dessen Natur, Entstehung und Wiedererzeugung. Leipzig.

Davis, J.B., V.F. Coty, and J.P. Stanley (1964) Atmospheric nitrogen fixation by methane-oxidizing bacteria. J. Bacteriol. 88:468-472.

Davis, M.B. (1968) Pollen grains in lake sediments: Redeposition caused by seasonal water circulation. Science 162:796-799.

Davis, R.B., and F. Berge (1980) Atmospheric deposition in Norway during the last 300 years as recorded in SNSF lake sediments. II. Diatom stratigraphy and inferred pH. Pages 270-271, Ecological Impact of Acid Precipitation. Proceedings of an International Conference, Sandefjord, Norway, March 11-14, 1980, edited by D. Drablös and A. Tollan. Oslo-Aas, Norway: SNSF project.

Davis, R.B., S.A. Norton, D.F. Brakke, F. Berge, and C.T. Hess (1980) Atmospheric deposition in Norway during the last 300 years as recorded in SNSF lake sediments. IV. Synthesis, comparison with New England. Pages 274-275, Ecologic Impact of Acid Precipitation. Proceedings of an International Conference, Sandefjord, Norway, March 11-14, 1980, edited by D. Drablös and A. Tollan. Oslo-Aas, Norway: SNSF project.

Davy, H. (1821) Elements of Agricultural Chemistry. 3rd ed. London: Longman, Hurst, Rees, Orme, and Brown.

Daye, P.G., and E.T. Garside (1979) Development and survival of embryos and alevins of the Atlantic Salmon, Salmo salar L., continuously exposed to acidic levels of pH from fertilization. Can. J. Zool. 57:1713-1718.

Daye, P.G., and E.T. Garside (1980) Structural alterations in embryos and alevins of the Atlantic salmon, Salmo salar L., induced by continuous or short-term exposure to acidic levels of pH. Can. J. Zool. 58:27-43.

Delwiche, C.C., and B.A. Bryan (1976) Denitrification. Annu. Rev. Microbiol. 30:241-262.

Delwiche, C.C., and G.E. Likens (1977) Biological response to fossil fuel combustion products. Pages 73-88, Global Chemical Cycles and their Alterations by Man, edited by W. Stumm. Physical and Chemical Research Report 2. Berlin: Dahlem Konferenzen.

Delwiche, C.C., S. Bissell, and R. Virginia (1978) Soil and other sources of nitrogen oxide. Pages 459-476, Nitrogen in the

Environment, edited by D.R. Nielsen and J.G. Macdonald. New York: Academic Press.

Denmead, O.T., J.R. Simpson, and J.R. Freney (1974) Ammonia flux into the atmosphere from a grazed pasture. Science 185:609-610.

Denmead, O.T., J.R. Freney, and J.R. Simpson (1976) A closed ammonia cycle within a plant canopy. Soil Biol. Biochem. 8:161-164.

Denmead, O.T. (1979) Chamber systems for measuring nitrous oxide emission from soils in the field. Soil Sci. Soc. Am. J. 43:89-95.

Denmead, O.T., J.R. Freney, and J.R. Simpson (1979) Nitrous oxide emission during denitrification in a flooded field. Soil Sci. Soc. Am. J. 43:716-718.

Dickson, W. (1975) The acidification of Swedish lakes. Res. Rep. Inst. Freshwater Res. Drottningholm 54:8-20.

Dickson, W. (1980) Properties of acidified waters. Pages 75-83, Ecological Impact of Acid Precipitation. Proceedings of an International Conference, Sandefjord, Norway, March 11-14, 1980, edited by D. Drablös and A. Tollan. Oslo-Aas, Norway: SNSF project.

Dillon, P.J., D.S. Jeffries, W. Snyder, R. Reid, N.D. Yan, D. Evans, J. Moss, and W.A. Scheider (1978) Acidic precipitation in south-central Ontario: Recent observations. J. Fish. Res. Bd. Can. 35:809-815.

Dillon, P.J., D.S. Jeffries, W.A. Scheider, and N.D. Yan (1980) Some aspects of acidification in southern Ontario. Pages 212-213, Ecological Impact of Acid Precipitation. Proceedings of an International Conference, Sandefjord, Norway, March 11-14, 1980, edited by D. Drablös and A. Tollan. Oslo-Aas, Norway: SNSF project.

Dochinger, L.S., F.W. Bender, F.O. Fox, and W.W. Heck (1970) Chlorotic dwarf of eastern white pine caused by ozone and sulfur dioxide interaction. Nature 225:476.

Dochinger, L.S., and T.A. Seliga (1976) Proceedings of the First International Symposium on Acid Precipitation and the Forest Ecosystem. General Technical Report NE-23. Upper Darby, PA: USDA, Forest Service.

Doran, J.W., and M. Alexander (1977a) Microbial transformation of selenium. Appl. Environ. Microbiol. 33:31-37.

Doran, J.W., and M. Alexander (1977b) Microbial formation of volatile selenium compounds in soil. Soil Sci. Soc. Am. J. 41:70-73.

Dovland, H., and A. Semb (1980) Atmospheric transport of pollutants. Pages 14-21, Ecological Impact of Acid Precipitation. Proceedings of an International Conference, Sandefjord, Norway, March 11-14, 1980, edited by D. Drablös and A. Tollan. Oslo-Aas, Norway: SNSF project.

Dowdell, R.J., and K.A. Smith (1974) Field studies of the soil atmosphere. II. Occurrence of nitrous oxide. J. Soil Sci. 25:231-238.

Drablös, D., and I.H. Sevaldrud (1980) Lake acidification, fish damage and utilization of outfields. A comparative survey of six highland areas, southeastern Norway. Pages 354-355, Ecological Impact of Acid Precipitation. Proceedings of an International

Conference, Sandefjord, Norway, March 11-14, 1980, edited by D. Drablös and A. Tollan. Oslo-Aas, Norway: SNSF project.

Drablös, D., and A. Tollan, eds. (1980) Ecological Impact of Acid Precipitation. Proceedings of an International Conference, Sandefjord, Norway, March 11-14, 1980. Oslo-Aas, Norway: SNSF project.

Drablös, D., I.H. Sevaldrud, and J.A. Timberlid (1980) Historical land-use changes related to fish status development in different areas in southern Norway. Pages 367-369, Ecological Impact of Acid Precipitation. Proceedings of an International Conference, Sandefjord, Norway, March 11-14, 1980, edited by D. Drablös and A. Tollan. Oslo-Aas, Norway: SNSF project.

Driscoll, C.T. (1980) Aqueous speciation of aluminum in the Adirondack region of New York State, USA. Pages 214-215, Ecological Impact of Acid Precipitation. Proceedings of an International Conference, Sandefjord, Norway, March 11-14, 1980, edited by D. Drablös and A. Tollan. Oslo-Aas, Norway: SNSF project.

Duce, R.A. (1978) Speculations on the budget of particulate and vapor phase non-methane organic carbon in the global troposphere. Pure Appl. Geophys. 116:244-274.

Duce, R.A., J.G. Quinn, C.E. Olney, S.R. Piotrowicz, B.J. Ray, and T.L. Wade (1972) Enrichment of heavy metals and organic compounds in the surface microlayer of Narragansett Bay, Rhode Island. Science 176:161-163.

Duce, R.A., G.L. Hoffman, and W.H. Zoller (1974) Atmospheric trace metals at remote northern and southern hemisphere sites: Pollution or natural? Science 187:59-61.

DuRietz, G.E. (1949) Huvudenheter och huvudgränser i svensk myrvegetation. Sven. Botanisk Tidskr. 43:274-309.

Dvorak, A.J., et al. (1978) Impacts of Coal-fired Power Plants on Fish, Wildlife, and Their Habitats. FWS/OBS-78/29. Argonne, IL: Argonne National Laboratory.

Eastman, J.A., and D.H. Stedman (1977) A fast response sensor for zone eddy correlation flux measurements. Atmos. Environ. 11:1209-1211.

Eaton, J.G. (1973) Chronic toxicity of a copper, cadmium and zinc mixture to the fathead minnow. (Pimephales promelas Rafinesque). Water Res. 7:1723-1736.

Edgington, D.N., S.A. Gordon, M.M. Thommes, and L.R. Almodovar (1970) The concentration of radium, thorium and uranium in tropical marine algae. Limnol. Oceanogr. 15:945-955.

Ehrenberg, C. (1849) Passatstaub und Blutregen. Pages 269-460, Phys. Abhandl. Konigl. Akad. Wiss. Berlin, 1847.

Eisler, R. (1973) Annotated bibliography on biological effects of metals in aquatic environments Ecological Research Series EPA-R3-73-007. National Environmental Research Center, Office of Research and Monitoring. Corvallis, OR: U.S. Environmental Protection Agency.

Eisler, R., and M. Wapner (1975) Second annotated bibliography on biological effects of metals in aquatic environments. Ecological Research Series EPA-600/3-75-008. Office of Research and

Development, Environmental Research Laboratory. Narragansett, RI: U.S. Environmental Protection Agency.

Elliott, L.F., G.E. Schuman, and F.G. Viets (1971) Volatilization of nitrogen-containing compounds from beef cattle areas. Soil Sci. Soc. Am. Proc. 35:752-755.

Elzerman, A.W., and D.E. Armstrong (1979) Enrichment of zinc, cadium, lead, and copper in the surface microlayer of Lakes Michigan, Ontario, and Mendota. Limnol. Oceanogr. 24:133-144.

Emanuelsson, A., E. Eriksson, and H. Egnér (1954) Composition of atmospheric precipitation in Sweden. Tellus 6:261-267.

Emerson, S., and R.H. Hesslein (1973) Distribution and uptake of artificially introduced Radium-226 in a small lake. J. Fish. Res. Bd. Can. 30:1485-1490.

Emerson, S. (1975a) Chemically-enhanced CO_2 gas exchange in an eutrophic lake: A general model. Limnol. Oceanogr. 20:743-753.

Emerson, S. (1975b) Gas exchange rates in small Canadian Shield lakes. Limnol. Oceanogr. 20:754-761.

Epps, E.A., and M.B. Sturgis (1939) Arsenic compounds toxic to rice. Soil Sci. Soc. Am. Proc. 4:215-218.

Eriksson, E. (1952) Composition of atmospheric precipitation. I. Nitrogen compounds. Tellus 4:215-232. II. Sulfur, chloride, iodine compounds. Bibliography. Tellus 4:280-303.

Eriksson, E. (1958) The chemical climate and saline soils in the arid zone. Arid Zone Res. 10:147-180.

Eriksson, E. (1959) The yearly circulation of chloride and sulfur in nature; meteorological, geochemical and pedological implications, Part I. Tellus 11:375-403.

Eriksson, E. (1960) The yearly circulation of chloride and sulfur in nature; meteorological, geochemical and pedological implications, Part II. Tellus 12:63-109.

European Inland Fisheries Advisory Commission (1969) Water quality criteria for European freshwater fish--extreme pH values and inland fisheries. Water Research 3:593-611.

Evans, L.S., and T.M. Curry (1979) Differential responses of plant foliage to simulated acid rain. Am. J. Botany 66:953-963.

Evans, L.S., C.A. Conway, and K.F. Lewin (1980) Yield responses of field-grown soybeans exposed to simulated acid rain. Pages 162-163, Ecological Impact of Acid Precipitation. Proceedings of an International Conference, Sandefjord, Norway, March 11-14, 1980, edited by D. Drablös and A. Tollan. Oslo-Aas, Norway: SNSF project.

Evelyn, J. (1661) Fumifugium. London: Bedel and Collins. (Reprinted edition of 1772 seen.)

Farmer, J.G., D.S. Swan, and M.S. Baxter (1980) Records and sources of metal pollutants in a dated Loch Lomond sediment core. Science of the Total Environment 16:131-147.

Farrell, E.P., I. Nilsson, C.O. Tamm, and G. Wiklander (1980) Effects of artificial acidification with sulfuric acid on soil chemistry in a Scots pine forest. Pages 186-187, Ecological Impact of Acid Precipitation. Proceedings of an International Conference,

Sandefjord, Norway, March 11-14, 1980, edited by D. Drablös and A. Tollan. Oslo-Aas, Norway: SNSF project.

Fenn, L.B., and D.E. Kissel (1974) Ammonium volatilization from surface applications of ammonium. Soil. Sci. Soc. Am. Proc. 38:606-610.

Ferguson, P., J.A. Lee, and J.N.B. Bell (1978) Effects of sulphur pollution on the growth of Sphagnum species. Environ. Pollut. 16:151-162.

Ferry, B.W., M.S. Baddeley, and D.L. Hawksworth, eds. (1973) Air Pollution and Lichens. London: Athlone Press.

Fitchko, J., and T.C. Hutchinson (1975) A comparative study of heavy metal concentrations in river mouth sediments around the Great Lakes. J. Great Lakes Res. 1:46-78.

Fleming, R.W., and M. Alexander (1972) Dimethyl selenide and dimethyl telluride formation by a strain of Penicillium. Appl. Microbiol. 24:424-429.

Flett, R.J, J.W.M. Rudd, and R.D. Hamilton (1975) Acetylene reduction assays for nitrogen fixation in freshwater: A note of caution. Appl. Microbiol. 29:580-583.

Flett, R.J., D.W. Schindler, R.D. Hamilton, and N.E.R. Campbell (1980) Nitrogen fixation in precambrian shield lakes. Can. J. Fish. Aquat. Sci. 37:494-505.

Flowers, E.C., R.A. McCormick, and K.R. Kurfis (1969) Atmospheric turbidity over the United States. J. Appl. Meteorol. 8:955-962.

Focht, D.D., and L.H. Stolzy (1978) Long-term denitrification studies in soils fertilized with $(^{15}NH_4)_2SO_4$. Soil Sci. Soc. Am. J. 42:894-898.

Fogg, T.R., and W.F. Fitzgerald (1979) Mercury in Southern New England Coastal Rains. J. Geophys. Res. 84:6987-6989.

Fowler, D. (1980) Wet and dry deposition of sulphur and nitrogen compounds from the atmosphere. Pages 9-27, Effects of Acid Precipitation on Terrestrial Ecosystems, edited by T.C. Hutchinson and M. Havas. New York: Plenum Press.

Francis, A.J., D. Olson, and R. Bernatsky (1980) Effect of acidity on microbial processes in a forest soil. Pages 166-167, Ecological Impact of Acid Precipitation. Proceedings of an International Conference, Sandefjord, Norway, March 11-14, 1980, edited by D. Drablös and A. Tollan. Oslo-Aas, Norway: SNSF project.

Freedman, B. (1978) Effects of Smelter Pollution near Sudbury, Ontario, Canada on Surrounding Forested Ecosystem. Ph.D. Thesis, Department of Botany, University of Toronto, Toronto.

Freedman, B., and T.C. Hutchinson (1980) Smelter pollution near Sudbury, Ontario, Canada and effects on forest litter decomposition. Pages 395-434, Effects of Acid Precipitation on Terrestrial Ecosystems, edited by T.C. Hutchinson and M. Havas. New York: Plenum Press.

Freney, J.R., O.T. Denmead, and J.R. Simpson (1978) Soil as a source or sink for atmospheric nitrous oxide. Nature 273:530-532.

Friberg, F., C. Otto, and B.S. Svensson (1980) Effects of acidification on the dynamics of allochthonous leaf material and benthic invertebrate communities in running waters. Pages 304-305,

Ecological Impact of Acid Precipitation. Proceedings of an International Conference, Sandefjord, Norway, March 11-14, 1980, edited by D. Drablös and A. Tollan. Oslo-Aas, Norway: SNSF project.

Friend, J.P. (1973) The global sulfur cycle. Pages 177-201, Chemistry of the Lower Atmosphere, edited by S.I. Rasool. New York: Plenum Press.

Fromm, P.O. (1980) A review of some physiological and toxicological responses of freshwater fish to acid stress. Environ. Biol. Fish. 5:79-93.

Gabriel, M.L., and S. Fogel, eds. (1955) Great Experiments in Biology. Englewood Cliffs, NJ: Prentice Hall.

Gächter, R., and W. Geiger (1979) MELIMEX, an experimental heavy metal pollution study: Behavior of heavy metals in an aquatic food chain. Schweiz. Z. Hydrol. 41:277-290.

Gächter, R., and A. Mares (1979) Effects of increased heavy metal load on phytoplankton communities. Schweiz. Z. Hydrol. 41:228-246.

Galbally, I.E. (1976) Emissions of oxides of nitrogen (NO_x) and ammonia from the earth's surface. Tellus 27:67-70.

Galloway, J.N., and E.B. Cowling (1978) The effects of precipitation on aquatic and terrestrial ecosystems--A proposed precipitation chemistry network. J. Air Pollut. Control Assoc. 28:229-235.

Galloway, J.M., and G.E. Likens (1979) Atmospheric enhancement of metal deposition in Adirondack lake sediments. Limnology and Oceanography 24:427-433.

Galloway, J.N., and D.M. Whelpdale (1980) An atmospheric sulfur budget for eastern North America. Atmos. Environ. 14:409-417.

Galloway, J.N., H.L. Volchok, D. Thornton, S.A. Norton, and R.A.N. McLean (1981) Toxic substances in atmospheric deposition: A review and assessment. Pages 19-82, The Potential Atmospheric Impact of Chemicals Released to the Environment, edited by J.M. Miller. EPA 560/5-80-001. Washington, DC: U.S. Environmental Protection Agency.

Ganther, H.E., P.A. Wagner, M.L. Sunde, and W.G. Hoekstra (1973) Protective effects of selenium against heavy metal toxicities. Pages 247-258, Trace Substances in Environmental Health, Vol. 6, edited by D.D. Hemphill. Columbia, MO: University of Missouri.

Ganther, H.E. (1978) Modification of methylmercury toxicity and metabolism by selenium and vitamin E.: Possible mechanisms. Environmental Health Perspectives 25:71-76.

Garcia, J.L. (1976) Nitric oxide production in rice soils. Ann. Microbiol. 127A:401-414.

Garland, J.A., W.S. Clough, and D. Fowler (1973) Deposition of sulfur dioxide on grass. Nature 242:256-257.

Garrels, R.M., F.T. Mackenzie, and C. Hunt (1975) Chemical Cycles and the Global Environment. Los Altos, CA: William Kaufmann Pub.

Geison, G.L. (1971) Ferdinand Julius Cohn. Dict. Sci. Biogr. 3:336-341.

Ghiorse, W.C., and M. Alexander (1976) Effect of microorganisms on the sorption and fate of sulfur dioxide and nitrogen dioxide in soil. J. Environ. Qual. 5:227-230.

Gibson, J. (1979) NADP News, August issue. Deposition Program. Natural Resources Ecology Laboratory, Colorado State University. Ft. Collins: National Atmospheric Deposition Program.

Gibson, J., and C.V. Baker (1979) NADP First Data Report, July 1978 through February 1979. Natural Resources Ecology Laboratory, Colorado State University. Ft. Collins: National Atmospheric Deposition Program.

Gjessing, E., A. Henriksen, M. Johannessen, and R.F. Wright (1976) Effects of acid precipitation on freshwater chemistry. Pages 65-86, Impact of Acid Precipitation on Forest and Freshwater Ecosystems in Norway, edited by F.H. Braekke. Research Report FR-6. NISK, Aas, Norway: SNSF Project.

Glass, G., and O.L. Loucks, eds. (1980) Impacts of Airborne Pollutants on Wilderness Areas along the Minnesota-Ontario border. Environmental Research Laboratory, Office of Research and Development. EPA 600/3-80-044. Duluth, MN: U.S. Environmental Protection Agency.

Goldberg, A.B., P.J. Maroulis, L.A. Wilner, and A.R. Bandy (1981) Study of H_2S emissions from a salt water marsh. J. Atmos. Environ. 15:11-18.

Goldberg, E.D. (1974) The surprise factor in marine pollution studies. J. Marine Technol. Soc. 8:29-34.

Goldberg, E.D. (1975) Synthetic organohalides in the sea. Proc. R. Soc. Lond. [Biol.] 189:277-289.

Goldberg, E.D. (1976) The Health of the Oceans. Paris: UNESCO Press.

Goldberg, E.D. (1979a) Proc. of a Workshop on Scientific Problems Relating to Ocean Pollution. Environmental Research Laboratories, National Oceanic and Atmospheric Administration. Washington, DC: U.S. Department of Commerce.

Goldberg, E.D. (1979b) Proc. of a Workshop on Assimilative Capacity of U.S. Coastal Waters for Pollutants. Environmental Research Laboratories, National Oceanic and Atmospheric Administration. Washington, DC: U.S. Department of Commerce.

Goldberg, E.D., V.F. Hodge, J.J. Griffin, and M. Koide (1981) The impact of fossil fuel combustion on the sediments of Lake Michigan. Environ. Sci. Technol. 15:446-471.

Gordon, A.G., and E. Gorham (1963) Ecological aspects of air pollution from an iron-sintering plant at Wawa, Ontario. Can. J. Botany 41:1063-1078.

Gorham, E. (1953) Some early ideas concerning the nature, origin and development of peat lands. J. Ecol. 41:257-274.

Gorham, E. (1954) An early view of the relation between plant distribution and environmental factors. Ecology 35:97-98.

Gorham, E. (1955) On the acidity and salinity of rain. Geochim. Cosmochim. Acta 7:231-239.

Gorham, E. (1958a) Accumulation of radioactive fallout by plants in the English Lake District. Nature 181:152-154.

Gorham, E. (1958b) Free acid in British soils. Nature 181:106.

Gorham, E. (1958c) Atmospheric pollution by hydrochloric acid. Q. J. R. Meteorol. Soc. 84:274-276.

Gorham, E. (1958d) The influence and importance of daily weather conditions in the supply of chloride, sulphate, and other ions to fresh waters from atmospheric precipitation. Philos. Trans. R. Soc. Lond. [Biol.] 247:147-178.

Gorham, E. (1959) A comparison of lower and higher plants as accumulators of radioactive fallout. Can. J. Botany 37:327-329.

Gorham, E., and J.B. Cragg (1960) The chemical composition of some bog waters from the Falkland Islands. J. Ecol. 48:175-181.

Gorham, E. (1961) Factors influencing supply of major ions to inland waters, with special reference to the atmosphere. Geol. Soc. Am. Bull. 72:795-840.

Gorham, E. (1965) Thomas Brotherton, Robert Hooke, and some neglected experiments in plant physiology during the late seventeenth century. BioScience 15:412.

Gorham, E. (1967) Some chemical aspects of wetland ecology. Pages 20-38, Tech. Mem. Assoc. Comm. Geotech. Res., Publication NRCC No. 80. Ottawa: National Research Council Canada.

Gorham, E. (1976) Acid precipitation and its influence upon aquatic ecosystems--An overview. Water, Air, Soil Pollut. 6:457-481.

Gorham, E., and J.E. Sanger (1976) Fossilized pigments as stratigraphic indicators of cultural eutrophication in Shagawa Lake, northeastern Minnesota. Geol. Soc. Am. Bull. 87:1638-1642.

Gorham, E. (1978a) Ecological aspects of the chemistry of atmospheric precipitation. Pages 265-296, Multidisciplinary Research Related to Atmospheric Sciences, edited by M.H. Glantz, H. van Loon, and E. Armstrong. NCAR/3141-78/1. Boulder, CO: National Center for Atmospheric Research.

Gorham, E. (1978b) The effects of acid precipitation upon aquatic and wetland ecosystems. Pages 37-45, A National Program for Assessing the Problem of Atmospheric Deposition (Acid Rain), edited by J.N. Galloway, E.B. Cowling, E. Gorham, and W.W. McFee. A report to the U.S. Council on Environmental Quality. Ft. Collins, CO: National Atmospheric Deposition Program.

Gorham, E., P.M. Vitousek, and W.A. Reiners (1979) The regulation of chemical budgets over the course of terrestrial ecosystem succession. Annu. Rev. Ecol. Syst. 10:53-84.

Gorham, E., and W.W. McFee (1980) Effects of acid deposition upon outputs from terrestrial to aquatic ecosystems. Pages 465-480, Effects of Acid Precipitation on Terrestrial Ecosystems, edited by T.C. Hutchinson and M. Havas. New York: Plenum Press.

Gosio, B. (1897) Zur Frage, wodurch die Giftigkeit arsenhaltiger Tapeten bedingt wird. Berichte 30:1024-1026.

Gough, L.P., H.T. Shacklette, and A.A. Case (1979) Element concentrations toxic to plants, animals and man. U.S. Geol. Survey Bull. 1466:80.

Graedel, T.E., I.A. Farrow, and T.A. Weber (1975) The influence of aerosols on the chemistry of the troposphere. Int. J. Chem. Kinet. Symp. 1:581.

Graedel, T.E. (1978) Chemical Compounds in the Atmosphere. New York: Academic Press.

Graham, B.M., R.D. Hamilton, and N.E.R. Campbell (1980) Comparison of the nitrogen-15 uptake and acetylene reduction methods for estimating the rates of nitrogen fixation by freshwater blue-green algae. Can. J. Fish. Aquat. Res. 37:488-493.

Grahn, O., H. Hultberg, and L. Landner (1974) Oligotrophication--A self-accelerating process in lakes subjected to excessive supply of acid substances. AMBIO 3:93-94.

Grahn, O. (1980) Fish kills in two moderately acid lakes due to high aluminum concentration. Pages 310-311, Ecological Impact of Acid Precipitation. Proceedings of an International Conference, Sandefjord, Norway, March 11-14, 1980, edited by D. Drablös and A. Tollan. Oslo-Aas, Norway: SNSF project.

Graunt, J. (1662) Natural and Political Observations Mentioned in a Following Index, and Made upon the Bills of Mortality. London: Martin, Allestry and Dicas.

Grice, G.D., and D.W. Menzel (1978) Controlled ecosystem pollution experiment: Effect of mercury on enclosed water columns. VIII. Summary of results. Mar. Sci. Commun. 4:23-33.

Griffin, J.J., and E.D. Goldberg (1979) Morphologies and origin of elemental carbon in the environment. Science 206:563-565.

Grosjean, D. (1977) Aerosols. Pages 45-125, Ozone and Other Photochemical Oxidants. Subcommittee on Ozone and Other Photochemical Oxidants, Committee on Medical and Biological Effects of Environmental Pollutants, Division of Medical Sciences, Assembly of Life Sciences. Washington, DC: National Academy of Sciences.

Grosjean, D., and S.K. Friedlander (1979) Formation of organic aerosols from cyclic olefins and diolefins. Pages 435-473, The Character and Origins of Smog Aerosols, edited by G.M. Hidy, P.K. Mueller, D. Grosjean, B.R. Appel, and J.J. Wesolowski. New York: John Wiley and Sons.

Groth, D.H., L.E. Stettler, and G. Mackay (1976) Interactions of mercury, cadmium, selenium, tellurium, arsenic and beryllium. Pages 527-543, Effects and Dose-Response Relationships of Toxic Metals, edited by G.F. Nordberg. Amsterdam: Elsevier.

Guerin, M.R. (1977) Energy Sources of Polycyclic Aromatic Hydrocarbons. Oak Ridge, TN: Oak Ridge National Laboratory.

Guerlac, H. (1954) The poets' nitre. Isis 45:243-255.

Haagen-Smit, A.J. (1952) Chemistry and physiology of Los Angeles smog. Ind. Eng. Chem. 44:1342-1346.

Haagen-Smit, A.J., C.E. Bradley, and M.M. Fox (1953) Ozone formation in photochemical oxidation of organic substances. Ind. Eng. Chem. 45:2086-2089.

Haagvar, S. (1980) Effects of acid precipitation on soil and forest. 7. Soil animals. Pages 202-203, Ecological Impact of Acid Precipitation. Proceedings of an International Conference, Sandefjord, Norway, March 11-14, 1980, edited by D. Drablös and A. Tollan. Oslo-Aas, Norway: SNSF project.

Häfele, W., and W. Sassin (1977) The global energy system. Pages 1-30, Annual Review of Energy, Vol. 2, edited by J.M. Hollander and M.K. Simmons. Palo Alto, CA: Annual Reviews.

Hales, S. (1738) Vegetable Staticks. 3rd ed. London: Innys and Manby, Woodward and Peele.

Hall, R.E., and D.G. DeAngelis (1980) EPA's research program for controlling residential wood combustion emissions. J. Air Pollut. Control Assoc. 30:862-867.

Halliday, E.C. (1961) A historical review of atmospheric pollution. Pages 9-37, Air Pollution. WHO Monograph Series, No. 46. Geneva: World Health Organization.

Hamilton, P.L. (1972) Aquarium experiments on the uptake of mercury by the chironomid Chironomus tentans Fabricius. Pages 89-92, Mercury in the Aquatic Environment, edited by J.F. Uthe. No. 1167. Fisheries Research Board of Canada, Manuscript Report Service. Winnipeg, Manitoba: Freshwater Inst., Dept. Supplies and Services.

Hansen, D.A., G.M. Hidy, and G.J. Stensland (1981) Examination of the basis for trend interpretation of historical rain chemistry in the eastern United States. Doc. No. P-A097. Westlake Village, CA: Environmental Research and Technology, Inc.

Hardy, J.T., and E.A. Crecelius (In press) Atmospheric particulate matter: Inhibition of marine primary productivity. Environ. Sci. Technol.Hardy, R.W.F., R.D. Holsten, E.K. Jackson, and R.C. Burns (1968) The acetylene-ethylene assay for measurement of nitrogen fixation: Laboratory and field evaluation. Plant Physiol. 43:1185-1207.

Harriss, R.C., and C. Hohenemser (1978) Mercury: Measuring and managing the risk. Environment 20(10):25-36.

Harriss, R.C., D.B. White, and R.B. Macfarlane (1979) Mercury compounds reduce photosynthesis by plankton. Science 170:736-737.

Harrison, H. (1970) Stratospheric ozone with added water vapor: Influence of high-altitude aircraft. Science 170:734-736.

Hartman, L.M. (1976) Fungal Flora of the Soil as Conditioned by Varying Concentrations of Heavy Metals. M.Sc. Thesis, University of Montana, Missoula, MT.

Harvey, H.H. (1980) Widespread and diverse changes in the biota of North American lakes and rivers coincident with acidification. Pages 93-98, Ecological Impact of Acid Precipitation. Proceedings of an International Conference, Sandefjord, Norway, March 11-14, 1980, edited by D. Drablös and A. Tollan. Oslo-Aas, Norway: SNSF project.

Harvey, W. (1628) De Motu Cordis, translated by C.D. Leake. 3rd ed., 1941. Philadelphia: Thomas.

Havas, M. (1980) A study of the Chemistry and Biota of Acid and Alkaline Ponds at the Smoking Hills. N.W. Territories. Ph.D. Thesis. Department of Botany. Toronto: University of Toronto.

Havlik, B. (1971) Radium in aquatic food chains: Radium uptake by fresh water algae. Radiat. Res. 46:490-505.

Hawksworth, D.L., S. Rose, and J.V. Coppins (1973) Changes in the lichen flora of England and Wales attributable to pollution of the air by sulfur dioxide. Pages 330-367, Air Pollution and Lichens, edited by B.W. Ferry, M.S. Baddeley, and D.L. Hawksworth. Toronto: University of London, Athalone Press.

Hawksworth, D.L., and S. Rose (1976) Lichens as Pollution Monitors. Southampton: Edward Arnold Publishers, Ltd.

Hawksworth, D.L., and M.R.D. Seaward (1977) Lichenology in the British Isles 1568-1975. Richmond, United Kingdom: Richmond Publishing Company Ltd.

Heichel, G.H. (1973) Removal of carbon monoxide by field and forest soils. J. Environ. Qual. 2:419-423.

Heidorn, K.C. (1978) A chronology of important events with history of air pollution meteorology to 1970. Bull. Am. Meteorol. Soc. 59: 1589-1597.

Hemond, H.F. (1980) Biogeochemistry of Thoreau's Bog, Concord, Massachusetts. Ecol. Monogr. 50:507-526.

Hemphill, D.D., and J.H. Rule (1975) Foliar uptake and translocation of Pb^{210} and Cd^{109} by plants. Pages 77-86, Proceedings of the International Conference on Heavy Metals in the Environment, Vol. 2, edited by T.C. Hutchinson. Toronto: University of Toronto, Institute for Environmental Studies.

Henderson, L.J. (1913) The Fitness of the Environment. New York: Macmillan (Reissued in 1958 with introduction by George Wald. Boston: Beacon Press).

Hendrey, G.R., K. Baalsrud, T.S. Traaen, M. Laake, and G. Raddum (1976) Acid precipitation: Some hydrobiological changes. AMBIO 5:224-227.

Henriksen, A. (1979) A simple approach for identifying and measuring acidification of freshwater. Nature 278:542-545.

Henriksen, A. (1980) Acidification of freshwaters--A large scale titration. Pages 68-74, Ecological Impact of Acid Precipitation. Proceedings of an International Conference, Sandefjord, Norway, March 11-14, 1980, edited by D. Drablös and A. Tollan. Oslo-Aas, Norway: SNSF project.

Henrikson, L., H.G. Oscarson, and J.A.E. Stenson (1980) Does the change of predator system contribute to the biotic development in acidified lakes? Page 316, Ecological Impact of Acid Precipitation. Proceedings of an International Conference, Sandefjord, Norway, March 11-14, 1980, edited by D. Drablös and A. Tollan. Oslo-Aas, Norway: SNSF project.

Herrmann, R., and J. Baron (1980) Aluminum mobilization in acid stream environments, Great Smoky Mountains National Park, USA. Pages 218-219, Ecological Impact of Acid Precipitation. Proceedings of an International Conference, Sandefjord, Norway, March 11-14, 1980, edited by D. Drablös and A. Tollan. Oslo-Aas, Norway: SNSF project.

Hesslein, R.H., W.S. Broecker, and D.W. Schindler (1980) Fates of metal radiotracers added to a whole lake: Sediment-water interactions. Can. J. Fish. Aquat. Res. 37:378-386.

Hetrick, F.M., M.D. Knittel, and J.L. Fryer (1979) Increased susceptibility of rainbow trout to infectious hematopoetic necrosis virus after exposure to copper. Appl. Environ. Microbiol. 37:198-201.

Hitchcock, D.R., L.L. Spiller, and W.E. Wilson (1980) Sulfuric acid aerosols and HCl release in coastal atmospheres: Evidence of rapid formation of sulfuric acid particulates. Atmos. Environ. 14:165-182.

Hodge, V.F., S.R. Johnson, and E.D. Goldberg (1978) Influence of atmospherically transported aerosols on surface ocean water composition. Geochem. J. 12:7-20.

Hodge, V.F., S.L. Seidel, and E.D. Goldberg (1979) Determination of tin(IV) and organotin compounds in natural waters, coastal sediments and macro algae by atomic absorption spectrometry. Analytical Chemistry 51:1256-1259.

Hoff, R.M., and A.J. Gallant (1980) Sulfur dioxide emissions from La Soufriere volcano, St. Vincent, West Indies. Science 209:923-924.

Hoffman, D.F., and J.M. Rosen (1980) Stratospheric sulfuric acid layer: Evidence for an anthropogeic component. Science 208:1368-1370.

Holdgate, M.W. (1979a) Targets of pollutants in the atmosphere. Philos. Trans. R. Soc. Lond. A 290:591-607.

Holdgate, M.W. (1979b) A Perspective of Environmental Pollution. Cambridge: Cambridge University Press.

Holmes, F.L. (1973) Justus von Liebig. Dict. Sci. Biogr. 8:329-350.

Holmes, J.A., E.C. Franklin, and R.A. Gould (1915) Report of the Selby Smelter Commission. U.S. Bureau of Mines Bulletin No. 98. Washington, DC: Government Printing Office.

Hooke, R. (1687) An account of several curious observations and experiments concerning the growth of trees. Philos. Trans. R. Soc. Lond. 16:307-313.

Horntvedt, R., G.J. Dollard, and E. Joranger (1980) Effects of acid precipitation on soil and forest. 2. Atmosphere-vegetation interactions. Pages 192-193, Ecological Impact of Acid Precipitation. Proceedings of an International Conference, Sandefjord, Norway, March 11-14, 1980, edited by D. Drablös and A. Tollan. Oslo-Aas, Norway: SNSF project.

Hosker, R.P., and S.E. Lindberg, eds. (In press) Atmospheric deposition and plant assimilation of gases and particles. Report of the Atmospheric Deposition and Uptake Panel of the Department of Energy's Second Ecology-Meteorology Workshop, University of Michigan Biological Experiment Station, Pellston, MI, Aug. 16-20, 1976. Atmos. Environ.

Houghton, H. (1955) On the chemical composition of fog and cloud water. J. Meteorol. 12:355-357.

Houston, D.B. (1974) Response of selected Pinus strobus L. clones to fumigations with sulfur dioxide and ozone. Can. J. Forest Res. 4:65-68.

Houston, D.B., and L.S. Dochinger (1977) Effects of ambient air pollutants on cone, seed and pollen characteristics in eastern white and red pine. Environ. Pollut. 12:1-5.

Howard, D.L., J.I. Frea, R.M. Pfister, and P.R. Dugan (1970) Biological nitrogen fixation in Lake Erie. Science 169:61-62.

Howarth, R.W. (1979) Pyrite: Its rapid formation in a salt marsh and its importance to ecosystem metabolism. Science 203:49-51.

Howarth, R.W., and J.M. Teal (1979) Sulfate reduction in a New England salt marsh. Limnol. Oceanogr. 24:999-1013.

Hultberg, H., and A. Wenblad (1980) Acid groundwater in southwestern
    Sweden. Pages 220-221, Ecological Impact of Acid Precipitation.
    Proceedings of an International Conference, Sandefjord, Norway, March
    11-14, 1980, edited by D. Drablös and A. Tollan. Oslo-Aas, Norway:
    SNSF project.

Humboldt, A., von (1805) Essai sur la geographie des plantes. Paris:
    Levrault and Schoell.

Husar, R.B., J.P. Lodge, and D.J. Moore, eds. (1978) Sulfur in the
    Atmosphere. Elmsford, NY: Pergamon Press.

Husar, R.B., D.E. Patterson, J.M. Holloway, W.E. Wilson, and T.G.
    Ellestad (1979) Trends of eastern U.S. haziness since 1948. Pages
    1-8, Proceedings of the 4th Symposium on Atmospheric Turbulence,
    Diffusion and Air Pollution, Reno, NV.

Hutchinson, G.E. (1954) The biochemistry of the terrestrial
    atmosphere. Pages 371-433, The Earth as a Planet, edited by G.P.
    Kuiper. Chicago: University of Chicago Press.

Hutchinson, G.E. (1957) A Treatise on Limnology. Vol. 1. Geography,
    Physics, and Chemistry. New York: John Wiley and Sons.

Hutchinson, T.C. (1973) Comparative studies of the toxicity of heavy
    metals to phytoplankton and their synergistic interactions. Water
    Pollut. Res. Can. 8:68-90.

Hutchinson, T.C., and H. Czyrska (1973) Cadmium and zinc toxicity and
    synergism to floating aquatic plants. Water Pollut. Res. Can.
    7:59-65.

Hutchinson, T.C., and P.M. Stokes (1975) Heavy metal toxicity and
    algal bioassay. Pages 340-343, Water Quality Criteria Special
    Technical Publication No. 573. Philadelphia: American Society of
    Testing and Materials.

Hutchinson, T.C., A. Fedorenko, J. Fitchko, A. Kuja, J. Van Loon, and
    J. Lichwa (1975) Movement and compartmentation of nickel and copper
    in an aquatic ecosystem. Pages 89-105, Trace Substances in
    Environmental Health, Vol. 11, edited by D.D. Hemphill. Columbia,
    MO: University of Missouri.

Hutchinson, T.C., and L.M. Whitby (1977) Effect of acid precipitation
    and heavy metal inputs on a boreal forest ecosystem, Sudbury,
    Ontario. Water, Air, Soil Pollut. 7:421-438.

Hutchinson, T.C., and F.W. Collins (1978) Effect of H^+ ion activity
    and Ca^{2+} on the toxicity of metals in the environment. Environ.
    Health Perspect. 25:47-52.

Hutchinson, T.C., W. Gizyn, M. Havas, and V. Zobens (1978) Effect of
    long-term lignite burns on arctic ecosystems at the Smoking Hills,
    N.W.T. Pages 317-332, Vol. XII, A symposium, J. Trace Substances
    in Environmental Health, edited by D.D. Hemphill. Columbia, MO:
    University of Missouri.

Hutchinson, T.C., W. Gizyn, M. Havas, and V. Zobens (1979a) Effect of
    long-term lignite burns on Arctic ecosystems at the Smoking Hills,
    N.W.T. Pages 317-332, Trace Substances in Environmental Health,
    Vol. 12, edited by D.D. Hemphill. Columbia, MO: University of
    Missouri.

Hutchinson, T.C., J.C. Hellebust, D. Mackay, D. Tam, and P. Kauss
(1979b) Relationship of hydrocarbon solubility to toxicity in
algae and to cellular membrane effects. Pages 541-547,
Proceedings of the Oil Spill Conference of the American Petroleum
Institute, U.S. EPA and U.S. Coast Guard, Los Angeles, CA, March
1979. Washington, DC: American Petroleum Institute.

Hutchinson, T.C., and M. Havas, eds. (1980) Effects of Acid
Precipitation on Terrestrial Ecosystems. New York: Plenum Press.

Hutchinson, T.C. (In press) Long Term Effect of Oil Spills. Report to
the Department of Indian and Northern Affairs, Canada. Ottawa:
Arctic Land Use Research Program.

Idso, S.B. (1980) The climatological significance of a doubling of the
earth's carbon dioxide concentration. Science 207:1462-1463. See
also letters and reply in Science 210:6-8.

Ingersoll, R.B., R.E. Inman, and W.R. Fisher (1974) Soils' potential
as a sink for atmospheric carbon monoxide. Tellus 26:151-159.

International Joint Commission (1977) New and Revised Great Lake Water
Quality Objectives. Vol. 2. Windsor, ON: International Joint
Commission.

International Joint Commission (1980) Pollution in the Great Lakes
Basin from Land Use Activities. Windsor, ON: International Joint
Commission.

Irving, P.M., and J.E. Miller (1980) Response of field-grown soybeans
to acid precipitation alone and in combination with sulfur
dioxide. Pages 170-171, Ecological Impact of Acid Precipitation.
Proceedings of an International Conference, Sandefjord, Norway,
March 11-14, 1980, edited by D. Drablös and A. Tollan. Oslo-Aas,
Norway: SNSF project.

Jackson, T.A., G. Kipphut, R.H. Hesslein, and D.W. Schindler (1980)
Experimental study of trace metal chemistry in soft-water lakes at
different pH levels. Can. J. Fish. Aquat. Sci. 37:387-420.

Jacobson, J.S. (1980) The influence of rainfall composition on the
yield and quality of agricultural crops. Pages 41-46, Ecological
Impact of Acid Precipitation. Proceedings of an International
Conference, Sandefjord, Norway, March 11-14, 1980, edited by D.
Drablös and A. Tollan. Oslo-Aas, Norway: SNSF project.

Jaechke, W., H-W Georgii, H.C. Malewski, and H. Malewski (1978)
Contributions of H_2S to the atmospheric sulfur cycle. Pure and
Appl. Geophys. 116:465-475.

Jamieson, T.F. (1856) On the action of the atmosphere upon newly-
deepened soil. J. R. Agric. Soc. Engl. 17:407-473.

Jefferies, D.S., and W.R. Snyder (1981) Atmospheric deposition of
heavy metals in central Ontario. Water, Air, Soil Pollut.
15:127-152.

Jernelov, A., and A-L Martin (1975) Ecological implications of metal
metabolism by microorganisms. Ann. Rev. of Microbiol. 29:61-71.

Joensuu, O.I. (1971) Fossil fuels as a source of mercury pollution.
Science 172:1027.

Johannes, R.E., J. Alberts, C. D'elia, R.A. Kinzie, L.R. Pomeroy, W.
Sottile, and W. Wiebe (1972) The metabolism of some coral reef

209

communities: A team study of nutrient and energy flux at
Eniwetok. Bio. Science 22:541-543.

Johannessen, M. (1980) Aluminum, a buffer in acidic waters? Pages
222-223, Ecological Impact of Acid Precipitation. Proceedings of
an International Conference, Sandefjord, Norway, March 11-14,
1980, edited by D. Drablös and A. Tollan. Oslo-Aas, Norway: SNSF
project.

Johnsen, I. (1980) Regional and local effects of air pollution, mainly
sulfur dioxide, on lichens and bryophytes in Denmark. Pages
133-140, Effects of Acid Precipitation on Terrestrial Ecosystems,
edited by T.C. Hutchinson and M. Havas. New York: Plenum Press.

Johnson, N.M. (1979) Acid rain: Neutralization within the Hubbard
Brook ecosystem and regional implications. Science 204:497-499.

Johnston, H.S. (1971) Reduction of stratospheric ozone by nitrogen
oxide catalysts from supersonic transport exhausts. Science
173:517-522.

Jones, K., and W.D.P. Steward (1969) Nitrogen turnover in marine and
brackish habitats. III. The production of extracellular nitrogen
by Calothrik scopolorum. J. Marine Biol. Assoc. U.K. 49:475-488.

Jordan, M.J., and M.P. Lechevalier (1975) Effects of zinc smelter
emissions on forest soil microflora. Can. J. Microbiol. 21:1855.

Jowett, D. (1958) Populations of Agrostis spp. tolerant of heavy
metals. Nature 182:816-817.

Kaakinen, J.W., R.M. Jorden, M.H. Lawasani, and R.E. West (1975)
Trace element behavior in coal-fired power plant. Environ. Sci.
Technol. 9:862-869.

Kagi, J.H.R., and M. Nordberg (eds.) (1979) Proceedings of the First
International Meeting on Metallothionein and other Low Molecular
Weight Metal-Binding Proteins. Basel: Birkhäuser Verlag.

Kanwisher, J. (1963) Effects of wind on CO_2 exchange across the sea
surface. J. Geophys. Res. 68:3921-3927.

Katz, M., ed. (1939) Effect of Sulphur Dioxide on Vegetation.
Publication No. 815. Ottawa: National Research Council Canada.

Katz, M. (1961) Some aspects of the physical and chemical nature of
air pollution. Pages 97-158, Air Pollution. WHO Monograph
Series, No. 46. Geneva: World Health Organization.

Keeling, C.D. (1970) Is carbon dioxide from fossil fuel changing man's
environment? Proc. Am. Philos. Soc. 114:10-17.

Keeling, C.D., and R.B. Bacastow (1977) Impact of industrial gases on
climate. Pages 72-95, Energy and Climate. Panel on Energy and
Climate, Geophysics Study Committee, Assembly of Mathematical and
Physical Sciences. Washington, DC: National Academy of Sciences.

Keller, W., J. Gunn, and N. Conroy (1980) Acidification impacts on
lakes in the Sudbury, Ontario, Canada area. Pages 228-229,
Ecological Impact of Acid Precipitation. Proceedings of an
International Conference, Sandefjord, Norway, March 11-14, 1980,
edited by D. Drablös and A. Tollan. Oslo-Aas, Norway: SNSF
project.

Kellogg, W.W. (1978) Review of man's impact on global climate.
Pages 64-81, Multidisciplinary Research Related to Atmospheric
Sciences, edited by M.H. Glantz, H. van Loon, and E. Armstrong.

NCAR/3141-78/1. Boulder, CO: National Center for Atmospheric Research.

Kellogg, W.W., and R. Schware (1981) Climate Change and Society. Boulder, CO: Westview Press.

Kelly, T.J., D.H. Stedman, J.A. Ritter, and R.B. Harvey (1980) Measurements of oxides of nitrogen and nitric acid in clean air. J. Geophys. Res. 85:7417-7425.

Kenaga, E.E. (1980) Correlation of bioconcentration factors of chemicals in aquatic and terrestrial organisms with their physical and chemical properties. Environ. Sci. Technol. 14:553-556.

Kennedy, L.A. (1981) The effects of lake acidification on embryonic development of the lake trout, Salvelinus mamycush. Pages 49-54, Proceedings of the 6th Annual Aquatic Toxicology Workshop. Fisheries and Marine Service Technical Report No. 975. Ottawa: Fisheries and Marine Service.

Kerr, R.A. (1979) Global pollution: Is the arctic haze actually industrial smog? Science 205:290-293.

Kettlewell, H.B.D. (1955) Selection experiments on industrial melanism in the Lepidoptera. Heredity 9:323-342.

Kettlewell, H.B.D. (1956) Further selection experiments on industrial melanism in the Lepidoptera. Heredity 10:287-301.

Kivinen, E. (1935) Über Elektrolytgehalt und Reaktion der Moorwasser. Maatalouskoelaitoksen Maatutkimusosasto Agrogeologisia Julkaisuja No. 38.

Kneese, A.V., R.U. Ayres, and R.C. d'Arge (1970) Economics and the Environment--A Materials Balance Approach. Baltimore: Johns Hopkins University Press.

Koide, M., and E.D. Goldberg (1971) Atmospheric sulfur and fossil fuel combustion. J. Geophys. Res. 76:6589-6596.

Koike, I., and S. Hattori (1978) Denitrification and ammonia formation in anaerobic coastal sediments. Appl. Environ. Microbiol. 35:278-282.

Komsta-Szumska, E., J. Chmielnicka, and J.K. Piotrowski (1976) The influence of selenium on binding of inorganic mercury by metallothionein in the kidney and liver of the rat. Biochemical Pharmacology 25:2539-2540.

Kramer, J.R. (1976) Geochemical and lithological factors in acid precipitation. Pages 611-618, Proceedings of the First International Symposium on Acid Precipitation and the Forest Ecosystem, edited by L.S. Dochinger and T.A. Seliga. General Technical Report NE-23. Upper Darby, PA.: USDA, Forest Service.

Kress, L.W., and J.M. Skelly (1977) The interaction of O_3, SO_2, and NO_2 and its effect on the growth of two forest tree species. Pages 135-155, Air Pollution and Its Impact on Agriculture. Cottrell Centennial Symposium, January 13-14, 1977. Stanislaus, CA: California State College.

Kwain, W.E. (1975) Effects of temperature on development and survival of rainbow trout, Salmo gairdneri, in acid waters. J. Fish Res. Board Can. 32:493-497.

Laflamme, R.E., and R.A. Hites (1978) The global distribution of polycyclic aromatic hydrocarbons in recent sediments. Geochim. Cosmochim. Acta 42:289-303.

Landa, E.R. (1979) The volatile loss of mercury from soils amended with methylmercury chloride. Soil Sci. 128:9-16.

Landsberg, H.E. (1954) Some observations of the pH of precipitation elements. Arch. Meteorol. Geophys. Bioklimatologie, Series A, 7:219-226.

Lang, C., and B. Lang-Dobler (1979) Oligochaetes and chironomial larvae in heavy metal loaded and control limnocorrals. Schweiz. Z. Hydrol. 41:271-276.

Lantzy, R.J., and F.T. Mackenzie (1979) Atmospheric trace metals: Global cycles and assessment of man's impact. Geochim. Cosmochim. Acta 43:511-525.

Lau, N.C., and R.J. Charlson (1977) On the discrepancy between background atmospheric ammonia gas measurements and the existence of acid sulfates as a dominant atmospheric aerosol. Atmos. Environ. 11:475-478.

Lauer, D.A., D.R. Bouldin, and S.D. Klausner (1976) Ammonia volatilization from dairy manure spread on the soil surface. J. Environ. Qual. 5:134-140.

Lave, L.B., and E.P. Seskin (1977) Air Pollution and Human Health. Resources for the Future. Baltimore, MD: Johns Hopkins University Press.

Lazarek, S. (1980) Cyanophytan mat communities in acidified lakes. Naturwissenschaften 67:97-98.

Lazrus, A.L., E. Lorange, and J.P. Lodge (1970) Lead and other metal ions in United States precipitation. Environ. Sci. Technol. 4:55-58.

Leber, A.P., and T.S. Miya (1976) A mechanism for cadmium- and zinc-induced tolerance to cadmium toxicity: Involvement of metallothionein. Toxicol. and Appl. Pharm. 37:403-414.

Levander, O.A., and L.C. Argrett (1969) Effects of arsenic, mercury, vallium and lead on selenium metabolism in rats. Toxicol. and Appl. Pharm. 14:308-314.

Levander, O. (1977) Metabolic interrelationships between arsenic and selenium. Environ. Health Perspect. 19:159.

Levine, J.S., R.S. Rogowski, G.L. Gregory, W.E. Howell, and J. Fishman (In press) Simultaneous measurements of NO_x, NO and O_3 production in a laboratory electric discharge: Atmospheric implications. Geophys. Res. Lett.

Lewis, B.G. (1976) Selenium in biological systems and its pathways for volatilization in higher plants. Pages 389-409, Environmental Biogeochemistry, Vol. 1, edited by J.O. Nriagu. Ann Arbor, MI: Ann Arbor Science.

Lidén, K. (1961) Cesium 137 burdens in Swedish Laplanders and reindeer. Acta Radiol. 56:237-240.

Lidén, K., and M. Gustafsson (1967) Relationship and seasonal variation in ^{137}Cs in lichen, reindeer and man in northern Sweden. Pages 193-208, Radioecological Concentration Processes, edited by B. Aaberg and F.P. Hungate. Oxford: Pergamon Press.

Liebl, K.H., and W. Seiler (1976) Carbon monoxide and hydrogen destruction at the soil surface. Pages 215-229, Microbial Production and Utilization of Gases, edited by H.G. Schlegel, G.

Gottschalk, and N. Pfennig. Göettingen, West Germany: Akademie der Wissenschaften.

Likens, G.E., F.H. Bormann, and N.M. Johnson (1972) Acid rain. Environment 14:33-40.

Likens, G.E., and F.H. Bormann (1974) Linkages between terrestrial and aquatic ecosystems. BioScience 24:447-456.

Likens, G.E. (1976) Acid precipitation. Chem. Eng. News 54(48):29-44.

Likens, G.E., F.H. Bormann, R.S. Pierce, J.S. Eaton, and N.M. Johnson (1977) Biogeochemistry of a Forested Ecosystem. New York: Springer-Verlag.

Liljestrand, H.M., and J.J. Morgan (1978) Chemical composition of acid precipitation in Pasadena, California. Environ. Sci. Technol. 12:1271-1273.

Lindberg, S.E. (1980) Mercury partitioning in a power plant plume and its influence on atmospheric removal mechanism. Atmos. Environ. 14:227-231.

Lindberg, S.E., and S.B. McLaughlin (In press) Air pollutants' interactions with vegetation: Research needs in data acquisition and interpretation. Pages __-__, Air Pollutants and Their Effects on the Terrestrial Ecosystem: Current Status and Future Needs of Research on Effects and Technology, edited by S.V. Krupa and A.H. Legge. New York: John Wiley and Sons.

Linzon, S.N. (1971) Economic effects of sulfur dioxide on forest growth. J. Air Pollut. Control Assoc. 21:81-86.

Lipfert, F.W. (1980) Sulfur oxides, particulates, and human mortality: Synopsis of statistical correlations. J. Air Pollut. Control Assoc. 30:366-371.

Liss, P., and P.G. Slater (1974) Flux of gases across the air-sea interface. Nature 247:181-184.

Liss, P. (1975) Chemistry of the air-sea interface. Pages 193-243, Chemical Oceanography, Vol. 2, 2nd ed., edited by J.P. Riley and G. Skirrow. London: Academic Press.

Llewellyn, R.A., and W.J. Washington (1977) Regional and global aspects. Pages 106-118, Energy and Climate. Panel on Energy and Climate, Geophysics Study Committee, Assembly of Mathematical and Physical Sciences. Washington, DC: National Academy of Sciences.

Lodge, J.P., Jr. (1980) An anecdotal history of air pollution. Advances Environ. Sci. Technol. 10:1-37.

Lohm, U. (1980) Effects of experimental acidification on soil organism populations and decomposition. Pages 178-179, Ecological Impact of Acid Precipitation. Proceedings of an International Conference, Sandefjord, Norway, March 11-14, 1980, edited by D. Drablös and A. Tollan. Oslo-Aas, Norway: SNSF project.

Loucks, D.L. (1979) Long-Term Ecological Research, Concept Statement and Measurement Needs. Summary of a Workshop. Indianapolis, IN: The Institute of Ecology.

Loughnan, F.C. (1969) Chemical Weathering of Silicate Minerals. New York: Elsevier.

Lovelock, J.E., R.J. Maggs, and R.A. Rasmussen (1972) Atmospheric dimethyl sulphide and the natural sulphur cycle. Nature 237:452-453.

Luebs, R.E., K.R. Davis, and A.E. Laag (1973) Enrichment of the atmosphere with nitrogen compounds volatilized from a large dairy area. J. Environ. Qual. 2:137-141.

Lunde, G., J. Gether, N. Gjos, and M.-B.S. Lande (1976) Organic micropollutants in precipitation in Norway. Research Report FR9/76. Oslo-Aas: SNSF Project.

Lutz, H.J., and R.F. Chandler, Jr. (1946) Forest Soils. New York: John Wiley and Sons.

MacCracken, M.C. (1978) MAP3S: An investigation of atmospheric energy related pollutants in the Northeastern United States. Atmos. Environ. 12:649-659.

MacIntyre, F. (1974) Chemical fractionation and sea-surface microlayer processes. Pages 245-249, The Sea, Vol. 5: Marine Chemistry, edited by E.D. Goldberg. New York: Wiley Interscience.

Macketeth, F.J.N. (1966) Some chemical observations on post-glacial lake sediments. Phil. Trans. Roy. Soc. Lond. (B) 250:165-213.

MacLean, D.C. (1975) Stickstoffoxide als phytotoxische Luftverunreinigungen. Staub Reinhaltung der Luft 35:205-210.

Mague, T.H., and O. Holm-Hansen (1975) Nitrogen fixation on a coral reef. Phycologia 14:87-92.

Makarov, Y.N. (1976) Systematic characteristics and distribution of larvae of Brachyura (Decapoda) in the Black Sea neuston (English summary). Zoologicheskii Zhurnal 55:363-370.

Malley, D.F. (1980) Decreased survival and calcium uptake by the crayfish Oronectes virilis in low pH. Can. J. Fish. Aquat. Sci. 37:364-372.

Malmer, N. (1974) On the effect on water, soil and vegetation of an increasing atmospheric supply of sulfur. Publication No. PM 402E. Solna, Sweden: National Swedish Environment Protection Board.

Malmer, N. (1976) Acid precipitation: Chemical changes in the soil. AMBIO 5:231-234.

Marchetti, C. (1975) Primary energy substitution model. On the interaction between energy and society. Chem. Econ. Eng. Rev. 1:9-15.

Maroulis, P.J., and A.R. Bandy (1977) Estimate of the contribution of biologically produced dimethyl sulfide to the global sulfur cycle. Science 196:647-648.

Marsh, A.R.W. (1978) Sulphur and nitrogen contributions to the acidity of rain. Atmos. Environ. 12:401-406.

Marshall, J.S., and D.L. Mellinger (1980) Dynamics of cadmium-stressed plankton communities. Can. J. Fish. Aquat. Sci. 37:403-414.

Matsunaga, K., and T. Goto (1976) Mercury in air and precipitation. Geochem. J. 10:107-109.

Mattson, S., G. Sandberg, and P.E. Terning (1944) Electrochemistry of soil formation. VI. Atmospheric salts in relation to soil and peat formation and plant composition. Kungl. Lantbrukshögskolans Ann. 12:101-118.

Mattson, S., and E. Koutler-Andersson (1954) Geochemistry of a raised bog. Kungl. Lantbrukshögskolans Ann. 21:321-366.

Mayer, R. (1979) Estimation of root uptake, biomass production and decomposition as sources or sinks for hydrogen ions. Section 4.3

(2 pages), Ecological Effects of Acid Precipitation, EPRI Report
No. EA-79-6-LD. Palo Alto, CA: Electric Power Research Institute.

McCarroll, J. (1980) Health effects associated with increased use of
coal. J. Air Pollut. Control Assoc. 30:652-656.

McConnell, K.P., and D.M. Carpenter (1971) Interrelationship between
selenium and specific trace elements. Proc. of the Soc. for
Experimental Biol. and Medicine 137:996-1001.

McCormick, R.A., and J.H. Ludwig (1967) Climate modification by
atmospheric aerosols. Science 156:1358-1360.

McCormick, J.N., G.N. Stokes, and G.J. Portele (1980) Environmental
acidification impact detection by examination of mature fish
ovaries. Pages 41-48, Proc. 6th Am. Aquat. Tox. Workshop, edited
by J.H. Klaverkamp, S.L. Leonhard, and K.E. Marshall. Can. Tech.
Rep. Fisheries and Aquatic Science No. 975. Winnipeg, Manitoba:
Western Region Dept. of Fisheries and Oceans.

McDonald, D.G., H. Hobe, and C.M. Wood (1980) The influence of calcium
on the physiological responses of the rainbow trout, Salmo
gairdneri, to low environmental pH. J. Exp. Biol. 88:109-131.

McFee, W.W. (1978) Effects of acid precipitation and atmospheric
deposition on soils. Pages 64-73, A National Program for
Assessing the Problem of Atmospheric Deposition (Acid Rain),
edited by J.N. Galloway, E.B. Cowling, E. Gorham, and W.W. McFee.
A Report to the U.S. Council on Environmental Quality. Ft.
Collins, CO: National Atmospheric Deposition Program.

McKim, J.M. (1977) Evaluation of tests with early life stages of fish
for predicting long-term toxicity. J. Fish. Res. Bd. Can.
34:1148-1154.

McKinnell, R.G., E. Gorham, F.B. Martin, and J.W. Schaad IV (1979)
Reduced prevalence of the Lucké renal adenocarcinoma in
populations of Rana pipiens in Minnesota. J. Natl. Cancer Inst.
63:821-824.

McLean, R.O. (1974) The tolerance of Stigeoclonium tenue Kütz to heavy
metals in South Wales. Br. Phycol. J. 9:91-95.

Mendelsohn, R., and G. Orcutt (1979) An empirical analysis of air
pollution dose-response curves. J. Environ. Econ. Manage.
6:85-106.

Menendez, R. (1976) Chronic effects of reduced pH on brook trout
(Salvelinus fontinalis). J. Fish. Res. Bd. Can. 33:118-123.

Menser, H.A., and H.E. Heggestad (1966) Ozone and sulfur dioxide
synergism: Injury to tobacco plants. Science 153:424-425.

Middleton, J.T., J.B. Kendrick, Jr., and H.W. Schwalm (1950) Injury to
herbaceous plants by smog or air pollution. Plant Dis. Rep.
34:245-252.

Middleton, J.T., E.F. Darley, and R.F. Brewer (1958) Damage to
vegetation from polluted atmospheres. J. Air Pollut. Control
Assoc. 8:9-15.

Middleton, J.T. (1961) Photochemical air pollution damage to plants.
Annu. Rev. Plant Physiol. 12:431-448.

Miettinen, J. (1969) Enrichment of radioactivity by Arctic ecosystems
in Finnish Lappland. Pages 23-31, Symposium on Radioecology,
edited by D.J. Nelson and R.C. Evans. Clearinghouse for Federal

Scientific and Technical Information, CONF-670503, Biology &
Medicine (TID-4500). Springfield, VA: National Bureau of
Standards.

Milbrink, G., and N. Johansson (1975) Some effects of acidification on
roe of roach, Rutilis rutilis L., and perch, Perca fluviatilis L.,
with special reference to the Avaa lake system in eastern Sweden.
Pages 52-62, Report #54. Drottningholm, Sweden: Institute for
Freshwater Research.

Miller, N.H.J. (1913) The composition of rain-water collected in the
Hebrides and in Iceland. J. Scott. Meteorol. Soc., Series 3
16(30):141-158.

Mitre Corporation (1979) National Environmental Impact Projection
No. 1. Report No. HCP/P-6119. Washington, DC: U.S. Department
of Energy.

Moffett, G.B., and J.D. Yarbrough (1972) The effect of DDT, toxaphene
and dieldrin on succinic dehydrogenase activity in insecticide
resistant and susceptible Gambusia affinis. J. Agric. Food Chem.
20:558-560.

Molina, M.J., and F.S. Rowland (1974) Stratospheric sink for
chlorofluoromethanes: Chlorine atom catalysed destruction of
ozone. Nature 249:810-812.

Morris, S.C., P.D. Moskowitz, W.A. Sevian, S. Silberstein, and
L.D. Hamilton (1979) Coal conversion technologies: Some health
and environmental effects. Science 206:654-662.

Mosaic (1979) Acid from the sky. Mosaic (National Science
Foundation) 10:35-40.

Mosier, A.R., C.E. Andre, and F.G. Viets, Jr. (1973) Identification of
aliphatic amines volatilized from a cattle feedlot. Environ. Sci.
Technol. 7:642-644.

Mount, D.I. (1973) Chronic effect of low pH on fathead minnow
survival, growth and reproduction. Water Res. 7:987-993.

Muhlbaier, J., and G.T. Tisue (1981) Cadmium in the southern basin of
Lake Michigan. Water, Air, Soil Pollut. 14:3-11.

Müller, G., G. Grimmer, and H. Böhnke (1977) Sedimentary record of
heavy metals and polycyclic hydrocarbons in Lake Constance.
Naturwissenschaften 64:427-431.

Müller, M.M., V. Sundman, and J. Skeyins (1980) Denitrification in low
pH spodosols and peats determined with the acetylene inhibition
method. Appl. Environ. Microbiol. 40:235-239.

Müller, P. (1980) Effects of artificial acidification on the growth of
periphyton. Can. J. Fish. Aquat. Sci. 37:355-363.

Muniz, I.P., and H. Leivestad (1980) Acidification--Effects on
freshwater fish. Pages 84-92, Ecological Impact of Acid
Precipitation. Proceedings of an International Conference,
Sandefjord, Norway, March 11-14, 1980, edited by D. Drablös and A.
Tollan. Oslo-Aas, Norway: SNSF project.

Munn, R.E. (1976) Atmospheric transport and diffusion on the regional
scale. J. Great Lakes Res. (Suppl. 1) 2:1-20.

Murphy, T.J., and C.P. Rzeszutko (1977) Precipitation inputs of PCBs
to Lake Michigan. J. of Great Lakes Research 3:305-312.

Naismith, J. (1807) Elements of Agriculture. London.

Nash, L.K. (1957) Plants and the atmosphere. Pages 325-436, Harvard Case Histories in Experimental Science, Vol. 2, edited by J.B. Conant and L.K. Nash. Cambridge, MA: Harvard University Press.

National Aeronautics and Space Administration (1981) Report of the NASA Working Group on Tropospheric Program Planning. NASA Reference Report No. 1062. Washington, DC.

National Research Council (1972) Particulate Polycyclic Organic Matter. Committee on Biological Effects of Atmospheric Pollutants. Washington, DC: National Academy of Sciences.

National Research Council (1975) Principles for Evaluating Chemicals in the Environment. Committee for the Working Conference on Principles of Protocols for Evaluating Chemicals in the Environment. Washington, DC: National Academy of Sciences.

National Research Council (1976a) Environmental Effects of Chlorofluoromethane Release. Committee on Impacts of Stratospheric Change, Assembly of Mathematical and Physical Sciences. Washington, DC: National Academy of Sciences.

National Research Council (1976b) Vapor-Phase Organic Pollutants: Volatile Hydrocarbons and Oxidation Products. Committee on Medical and Biological Effects of Environmental Pollutants, Division of Medical Sciences, Assembly of Life Sciences. Washington, DC: National Academy of Sciences.

National Research Council (1977a) Energy and Climate. Panel on Energy and Climate, Geophysics Study Committee, Assembly of Mathematical and Physical Sciences. Washington, DC: National Academy of Sciences.

National Research Council (1977b) Ozone and other Photochemical Oxidants. Subcommittee on Ozone and Other Photochemical Oxidants, Committee on Medical and Biological Effects of Environmental Pollutants, Division of Medical Sciences, Assembly of Life Sciences. Washington, DC: National Academy of Sciences.

National Research Council (1977c) Nitrogen Oxides. Committee on Medical and Biologic Effects of Environmental Pollutants, Division of Medical Sciences, Assembly of Life Sciences. Washington, DC: National Academy of Sciences.

National Research Council (1977d) Carbon Monoxide. Committee on Medical and Biologic Effects of Environmental Pollutants, Division of Medical Sciences, Assembly of Life Sciences. Washington, DC: National Academy of Sciences.

National Research Council (1978a) Sulfur Oxides. Committee on Sulfur Oxides, Board on Toxicology and Environmental Health Hazards, Assembly of Life Sciences. Washington, DC: National Academy of Sciences.

National Research Council (1978b) The Tropospheric Transport of Pollutants and Other Substances to the Oceans, Workshop on Tropospheric Transport of Pollutants to the Ocean Steering Committee, Ocean Sciences Board, Assembly of Mathematical and Physical Sciences. Washington, DC: National Academy of Sciences.

National Research Council (1978c) An Assessment of Mercury in the Environment. Panel on Mercury of the Coordinating Committee for

Scientific and Technical Assessments of Environmental Pollutants, Environmental Studies Board, Commission on Natural Resources. Washington, DC: National Academy of Sciences.

National Research Council (1978d) Nitrates: An Environmental Assessment. Panel on Nitrates of the Coordinating Committee for Scientific and Technical Assessments of Environmental Pollutants, Environmental Studies Board, Commission on Natural Resources. Washington, DC: National Academy of Sciences.

National Research Council (1979a) Airborne Particles. Baltimore, MD: University Park Press.

National Research Council (1979b) Energy in Transition 1985-2010. Final report of the Committee on Nuclear and Alternative Energy Systems, National Research Council, National Academy of Sciences. San Francisco: W.H. Freeman and Company.

National Research Council (1980a) Lead in the Human Environment. Committee on Lead in the Human Environment, Environmental Studies Board, Commission on Natural Resources. Washington, DC: National Academy Press.

National Research Council (1980b) The International Mussel Watch. Report of a Workshop Sponsored by the Environmental Studies Board, Commission on Natural Resources. Washington, DC: National Academy Press.

National Research Council (1981) Testing for Effects of Chemicals on Ecosystems. Committee to Review Methods for Ecotoxicology, Environmental Studies Board, Commission on Natural Resources. Washington, DC: National Academy Press.

National Research Council (In press) Disel Cars: Risks, Benefits and Public Policy. Diesel Impacts Study Committee, Assembly of Engineering. Washington, DC: National Academy Press.

National Research Council Canada (1979) Effects of Mercury in the Canadian Environment. Publication NRCC No. 16739. Ottawa, Canada: National Research Council Canada.

National Science Foundation (1979) Organic Panel Draft Report, 1979. Atmospheric Chemistry Workshop, National Center for Atmospheric Research, Boulder, CO, October 16-21, 1978. Washington, DC: National Science Foundation.

Nelson, D.J., and R.C. Evans, eds. (1969) Symposium on Radioecology. Clearinghouse for Federal Scientific and Technical Information, CONF-670503, Biology & Medicine (TID-4500). Springfield, VA: National Bureau of Standards.

Nguyen, B.C., A. Gaudry, B. Bonfang, and G. Lambert (1978) Reevaluation of the role of dimethylsulfide in the sulfur budget. Nature 275:637-839.

Nordberg, M. (1978) Studies in metallothionien and cadmium. Environ. Res. 15:381-404.

Norton, S.A. (1976) Changes in chemical processes in soils caused by acid precipitation. Pages 711-724, Proceedings of the First International Symposium on Acid Precipitation and the Forest Ecosystem, edited by L.S. Dochinger and T.A. Seliga. General Technical Report NE-23. Upper Darby, PA: USDA, Forest Service.

Norton, S.A., R.F. Dubiel, D.R. Sasserulle, and R.B. Davis (1978) Paleolimnologic evidence for increased zinc loading in lakes of New England, U.S.A. Verh. Int. Ver. Limnol. 20:538-545.

Norton, S.A., and C.T. Hess (1980) Atmospheric deposition in Norway during the last 300 years as recorded in SNSF lake sediments. I. Sediment dating and chemical stratigraphy. Pages 268-269, Ecological Impact of Acid Precipitation. Proceedings of an International Conference, Sandefjord, Norway, March 11-14, 1980, edited by D. Drablös and A. Tollan. Oslo-Aas, Norway: SNSF project.

de Noyelles, F., Jr., R. Knoechel, D. Reinke, D. Treanor, and C. Altenhofen (1980) Continuous culturing of natural phytoplankton communities in the Experimental Lakes Area: Effects of enclosure, in situ incubation, light, phosphorus and cadmium. Can. J. Fish. Aquat. Sci. 37:424-433.

Nozhevnikova, A.N., and L.N. Yurganov (1978) Microbial aspects of regulating the carbon monoxide in the earth's atmosphere. Adv. Microbiol. Ecol. 2:203-204.

Nriagu, J.O. (1978) Deteriorative effects of sulfur pollution on materials. Pages 1-59, Sulfur in the Environment (part II): Ecological Impacts. New York: John Wiley and Sons.

Nriagu, J.O. (1979) Global inventory of natural and anthropogenic emissions of trace metals to the atmosphere. Nature 279:409-411.

Nyborg, M. (1978) Sulfur pollution and soils. Pages 359-390, Sulfur in the Environment, edited by J.O. Nriagu. New York: John Wiley and Sons.

Nyborg, M., and P.B. Hoyt (1978) Effect of soil acidity and liming on mineralization of soil nitrogen. Can. J. Soil Sci. 58:331-338.

Odén, S. (1964) Aspects of the atmospheric corrosion climate. Pages 103-114, Current Corrosion Research in Scandinavia.

Odén, S. (1967) Dagens Nyheter, 24 October. Stockholm.

Odén, S. (1968) The acidification of air and precipitation and its consequences in the natural environment. Ecology Committee Bulletin No. 1, National Science Research Council, Stockholm. Arlington, VA: Translation Consultants Ltd. Tr-1172.

Odén, S., and T. Ahl (1970) Försurningen av Skandinaviska Vatten. [The Acidification of Scandinavian Lakes and Rivers]. Ärsbok, Sweden: Särtryk ur Ymer.

Odén, S. (1976) The acidity problem--an outline of concepts. Pages 1-36, Proceedings of the First International Symposium on Acid Precipitation and the Forest Ecosystem. General Technical Report NE-23. Upper Darby, PA: USDA, Forest Service.

Odén, S. (1980) On experiences with acid rain in Europe. Testimony before the House of Representatives Subcommittee on Oversight and Investigation, 26 Feb. 96th Congress, 2nd session.

Ökland, J. (1980) Environment and snails (Gastropoda): Studies of 1,000 lakes in Norway. Pages 322-323, Ecological Impact of Acid Precipitation. Proceedings of an International Conference, Sandefjord, Norway, March 11-14, 1980, edited by D. Drablös and A. Tollan. Oslo-Aas, Norway: SNSF project.

Ökland, J., and K.A. Ökland (1980) pH level and food organisms for fish: Studies of 1,000 lakes in Norway. Pages 326-327,

Ecological Impact of Acid Precipitation. Proceedings of an International Conference, Sandefjord, Norway, March 11-14, 1980, edited by D. Drablös and A. Tollan. Oslo-Aas, Norway: SNSF project.

Ökland, K.A. (1980) Mussels and crustaceans: Studies of 1,000 lakes in Norway. Pages 324-325, Ecological Impact of Acid Precipitation. Proceedings of an International Conference, Sandefjord, Norway, March 11-14, 1980, edited by D. Drablös and A. Tollan. Oslo-Aas, Norway: SNSF project.

Ontario Ministry of the Environment (1978) Extensive monitoring of lakes in the greater Sudbury area, 1974-1978. Rexdale, ON: Ontario Ministry of the Environment.

Overrein, L.N., H.M. Seip, and A. Tollan (1980) Acid Precipitation--Effects on Forest and Fish, report FR 19/80. Oslo-Aas, Norway: SNSF Project.

Parizek, J. (1978) Interactions between selenium compounds and those of mercury or cadmium. Environ. Health Perspect. 25:53.

Parker, A. (1955) Report on the Investigation of Atmospheric Pollution. Report No. 27. Department of Scientific and Industrial Research. London: Her Majesty's Stationery Office.

Patrick, R., V.P. Binetti, and S.G. Halterman (1981) Acid lakes from natural and anthropogenic causes. Science 211:446-448.

Perhac, R.N. (1978) Sulfate regional experiment in the northeastern United States: The SURE program. Atmos. Environ. 12:641-647.

Pitts, J.N., Jr., K.A. Van Cauwenberghe, D. Grosjean, J.P. Schmid, D.R. Fitz, W.L. Belser, G.B. Knudsen, and P.M. Hynds (1978) Atmospheric reactions of polycyclic aromatic hydrocarbons: Facile formations of mutagenic nitroderivatives. Science 202:515-519.

Plass, G.N. (1956) The carbon dioxide theory of climatic change. Tellus 8:140-154.

Poundstone, W.N. (1980) Testimony on "acid rain" issues to the U.S. Senate Committee on Energy and Natural Resources, 28 May. 96th Congress, 2nd session. (issued separately as "Some Facts about Acid Rain." Washington, DC: National Coal Association).

Prather, R.J., S. Miyamoto, and H.L. Bohn (1973a) Nitric oxide sorption by calcareous soils: I. Capacity, rate, and sorption products in air dry soils. Soil Sci. Soc. Am. Proc. 37:877-879.

Prather, R.J., S. Miyamoto, and H.L. Bohn (1973b) Sorption of nitrogen dioxide by calcareous soils. Soil Sci. Soc. Am. Proc. 37:860-863.

Priestley, J. (1772) Observations on different kinds of air. Philos. Trans. R. Soc. Lond. 62:147-252.

Probst, G.S., W.F. Bousquet, and T.S. Miya (1977) Correlation of hepatic metallothionein concentrations with acute cadmium toxicity in the mouse. Toxicol. and Appl. Pharm. 39:61-69.

Prohaskan, J.R., and H.E. Ganther (1977) Interactions between selenium and methylmercury in rat brain. Chem. Biol. Interactions 16:155-167.

Raddum, G.G., A. Hobaek, E.R. Lömsland, and T. Johnsen (1980) Phytoplankton and zooplankton in acidified lakes in South Norway. Pages 332-333, Ecological Impact of Acid Precipitation. Proceedings of an International Conference, Sandefjord, Norway,

March 11-14, 1980, edited by D. Drablös and A. Tollan. Oslo-Aas, Norway: SNSF project.

Radke, L.F., J.L. Stith, D.A. Hegg, and P.V. Hobbs (1978) Airborne studies of particles and gases from forest fires. J. Air Pollut. Control Assoc. 28:30-34.

Rafinesque, C.S. (1819) Thoughts on atmospheric dust. Am. J. Sci. 1:397-400.

Rafinesque, C.S. (1820) Western Minerva, or American Annals of Knowledge and Literature. Single copy published at Lexington, KY, by Thomas Smith.

Rankama, K., and T.G. Sahama (1950) Geochemistry. Chicago: University of Chicago Press.

Rasmussen, R.A. (1974) Emission of biogenic hydrogen sulfide. Tellus 26:254-260.

Rebsdorf, A. (1980) Acidification of Danish soft-water lakes. Pages 238-239, Ecological Impact of Acid Precipitation. Proceedings of an International Conference, Sandefjord, Norway, March 11-14, 1980, edited by D. Drablös and A. Tollan. Oslo-Aas, Norway: SNSF project.

Reddy, K.R., W.H. Patrick, Jr., and R.E. Phillips (1978) The role of nitrate diffusion in determining the order and rate of denitrification in flooded soil. I. Experimental results. J. Soil Sci. Soc. Am. 42:268-278.

Reinert, R.A., A.S. Heagle, and W.W. Heck (1975) Plant responses to pollutant combinations. Pages 159-178, Responses of Plants to Air Pollution, edited by J.B. Mudd and T.E. Kozlowski. New York: Academic Press.

Renwick, J.A.A., and J. Potter (1981) Effects of sulfur dioxide on volatile terpene emissions from balsam fir. J. Air Pollut. Control Assoc. 31:65-68.

Reuss, J.O. (1975a) Chemical/biological relationships relevant to ecological effects of acid rainfall. EPA-660/3-75-032. Washington, DC: U.S. EPA Ecological Research Service.

Reuss, J.O. (1975b) Sulfur in the soil system. Pages 51-61, Sulfur in the Environment. St. Louis, MO: Missouri Botanical Garden.

Rhoads, D.C. (1973) The influence of deposit-feeding benthos on water turbidity and nutrient recycling. Am. J. Sci. 273:1-22.

Risebrough, R.W. (1972) Reply to Hazeltine. Nature 240:164.

Risser, P.G., and K.D. Cornelison (1979) Man and the Biosphere, U.S. Information Synthesis Project MAB-8 Biosphere Reserves. Norman, OK: University of Oklahoma Press.

Robbins, J.A., and D.N. Edgington (1977) The distribution of selected chemical elements in the sediments of southern Lake Michigan. Report No. ANL-76-88, Part III. Argonne, IL: Argonne National Laboratory.

Robinson, E., and R.C. Robbins (1968) Sources, Abundance and Fate of Gaseous Atmospheric Pollutants. New York: American Petroleum Institute.

Robinson, E., and R.C. Robbins (1970a) Gaseous nitrogen compound pollutants from urban and natural sources. J. Air Pollut. Control Assoc. 20:303-306.

Robinson, E., and R.C. Robbins (1971) Final Report, API SRI Project SCC-8507. Palo Alto, CA: Stanford Research Institute.

Robinson, G.D. (1977) Effluents of energy production: Particulates. Pages 61-71, Energy and Climate. Panel on Energy and Climate, Geophysics Research Board, Assembly of Mathematical and Physical Sciences. Washington, DC: National Academy of Sciences.

Rodgers, G.A. (1978) Dry deposition of atmospheric ammonia at Rothamsted in 1976 and 1977. J. Agric. Sci. (Cambridge) 90:537-542.

Rodhe, H. (1978) Budgets and turnover times of atmospheric sulfur compounds. Atmos. Environ. 12:671-680.

Rodhe, H., R. Soderlund, and J. Ekstedt (1980) Deposition of airborne pollutants on the Baltic. AMBIO 9:168-173.

Rodhe, W. (1949) The ionic composition of lake waters. Verh. Int. Ver. Limnol. 10:377-386.

Rogers, R.D. (1976) Methylation of mercury in agricultural soils. J. Environ. Qual. 5:454-458.

Rogers, R.D. (1979) Volatility of mercury from soils amended with various mercury compounds. Soil Sci. Soc. Am. J. 43:298-291.

Rogers, R.D., and J.C. McFarlane (1979) Factors influencing the volatilization of mercury from soil. J. Environ. Qual. 8:255-250.

Romell, L.G. (1932) Mull and duff as biotic equilibria. Soil Sci. 34:161-188.

Romell, L.G. (1935) Ecological problems of the humus layer in the forest. Memorandum 170. Ithaca, NY: Cornell Agricultural Experiment Station.

Rorison, I.H. (1980) The effects of soil acidity on nutrient availability and plant response. Pages 283-304, Effects of Acid Precipitation on Terrestrial Ecosystems, edited by T.C. Hutchinson and M. Havas. New York: Plenum Press.

Rosenqvist, I.T. (1978a) Alternative sources for acidification of river waters in Norway. Sci. Total Environ. 10:39-49.

Rosenqvist, I.T. (1978b) Acid precipitation and other possible sources for acidification of rivers and lakes. Sci. Total Environ. 10:271-272.

Rosseland, B.O. (1980) Physiological responses to acid water in fish. 2. Effects of acid water on metabolism and gill ventilation in brown trout, Salmo trutta L., and brook trout, Salvelinus fontinalis Mitchill. Pages 348-349, Ecological Impact of Acid Precipitation. Proceedings of an International Conference, Sandefjord, Norway, March 11-14, 1980, edited by D. Drablös and A. Tollan. Oslo-Aas, Norway: SNSF project.

Roustan, J.L., A. Aumaitre, and E. Salmon-Legagneur (1977) Characterization of malodours during anaerobic storage of pig wastes. Agric. Environ. 3:147-157.

Rudd, J.W.M., and R.D. Hamilton (1975) Factors controlling rates of methane oxidation and the distribution of methane oxidizers in a small, stratified lake. Arch. Hydrobiol. 75:522-538.

Rudd, J.W.M., A. Furutam, R.J. Flett, and R.D. Hamilton (1976) Factors controlling methane oxidation in shield lakes: The role of nitrogen fixation and oxygen concentration. Limnol. Oceanogr. 21:357-364.

Rudd, J.W.M., A. Furutam, R.J. Flett, and R.D. Hamilton (1976) Factors controlling methane oxidation in shield lakes: The role of nitrogen fixation and oxygen concentration. Limnol. Oceanogr. 21:357-364.

Rudd, J.W.M., M.A. Turner, B.E. Townsend, A. Swick, and A. Furutoni (1980) Dynamics of selenium in mercury-contaminated experimental freshwater ecosystems. Can. J. Fish. Aquat. Sci. 37:848-857.

Ruhling, A., and G. Tyler (1968) An ecological approach to the lead problem. Botaniska Notiser 121:321-342.

Ruhling, A., and G. Tyler (1970) Sorption and retention of heavy metals in the woodland moss, Hylocomnium splendens (Hedw.) Bret. et Sch. Oikos 21:92-97.

Ruhling, A., and G. Tyler (1971) Regional differences in the deposition of heavy metals over Scandinavia. J. Appl. Ecol. 8:497-507.

Russel, Sir E.J., and E.W. Russell (1950) Soil Conditions and Plant Growth. 8th ed. London: Longmans Green and Co.

Russell, E.W. (1973) Soil Conditions and Plant Growth. 10th ed. London: Longmans Green and Co.

Salisbury, E.J. (1922) The soils of Blakeney Point: A study of soil reaction and succession in relation to the plant covering. Ann. Botany 36:391-431.

Salisbury, E.J. (1925) Note on the edaphic succession in some dune soils with special reference to the time factor. J. Ecol. 13:322-328.

Sandholm, M. (1974) Selenium carrier proteins in mouse plasma. Acta Pharma. et Toxicol. 35:424-428.

Schindler, D.W. (1971) Light, temperature and oxygen regimes of selected lakes in the experimental lakes area, northwest Ontario. J. Fish. Res. Bd. Can. 28:157-169.

Schindler, D.W. (1977) Evolution of phosphorus limitation in lakes. Science 195:260-262.

Schindler, D.W., R.H. Hesslein, and G. Kipphut (1977) Interactions between sediments and overlying waters in an experimentally eutrophied Precambrian Shield lake. Pages 235-243, Interactions Between Sediment and Fresh Water, edited by H.L. Golterman. Proceedings of a Symposium, Amsterdam, Sept. 1976. The Hague: Wageningen.

Schindler, D.W. (1980a) Experimental acidification of a whole lake: A test of the oligotrophication hypothesis. Pages 370-374, Ecological Impact of Acid Precipitation. Proceedings of an International Conference, Sandefjord, Norway, March 11-14, 1980, edited by D. Drablös and A. Tollan. Oslo-Aas, Norway: SNSF project.

Schindler, D.W. (1980b) Evolution of the Experimental Lakes Project. Can. J. Fish. Aquat. Res. 37:313-319.

Schindler, D.W., R. Wagemann, R.B. Cook, T. Ruszczynski, and J. Prokopowich (1980a) Experimental acidification of Lake 223, experimental lake area: Background data and the first three years of acidification. Can. J. Fish. Aquat. Sci. 37:342-354.

Schindler, D.W., R.H. Hesslein, R. Wagimann, and W.S. Broecker (1980b) Effects of acidification on mobilization of heavy metals and radionuclides from the sediments of a freshwater lake. Can. J. Fish. Aquat. Sci. 37:373-377.

Schlesinger, R.B. (1979) Natural removal mechanisms for chemical pollutants in the environment. BioScience 29:95-101.

Schnitzer, M. (1980) Effect of low pH on the chemical structure and reactions of humic substances. Pages 203-222, Effects of Acid Precipitation on Terrestrial Ecosystems, edited by T.C. Hutchinson and M. Havas. New York: Plenum Press.

Schofield, C.L. (1976) Acid precipitation: Effects on fish. AMBIO 5:228-230.

Schofield, C.L. (1980) Processes limiting fish populations in acidified lakes. Pages 345-356, Atmospheric Sulfur Deposition—Environmental Impact and Health Effects, edited by D.S. Shriner, C.R. Richmond, and S.E. Lindberg. Ann Arbor, MI: Ann Arbor Science.

Schreiber, R.W. (1980) The brown pelican: An endangered species? BioScience 30:742-747.

Scott, W.D. (1978) The pH of cloud water and the production of sulfate. Atmos. Environ. 12:917-921.

Seidel, S.L., U.F. Hodge, and E.D. Goldberg (In press) Tin as an environmental pollutant. Thalassia Jugoslavica.

Seiler, W., and P.J. Crutzen (1980) Estimates of gross and net fluxes of carbon between the biosphere and the atmosphere from biomass burning. Climatic Change 2:207-248.

Seip, H.M. (1980) Acidification of freshwater—Sources and mechanisms. Pages 358-366, Ecological Impact of Acid Precipitation. Proceedings of an International Conference, Sandefjord, Norway, March 11-14, 1980, edited by D. Drablös and A. Tollan. Oslo-Aas, Norway: SNSF project.

Semb, A. (1978) Deposition of Trace Elements from the Atmosphere in Norway. FR 13/78. Oslo-Aas, Norway: SNSF Project.

Settle, D.M., and C.C. Patterson (1980) Lead in albacore: Guide to lead pollution in Americans. Science 207:1167-1176.

Shaw, Sir N., and J.S. Owens (1925) The Smoke Problem of Great Cities. London: Constable.

Shinn, J.N., and S. Lynn (1979) Do man-made sources affect the sulfur cycle of northeastern states? Environ. Sci. Technol. 13:1062-1067.

Shriner, D.S., and E.B. Cowling (1980) Effects of rainfall acidification on plant pathogens. Pages 435-442, Effects of Acid Precipitation on Terrestrial Ecosystems, edited by T.C. Hutchinson and M. Havas. New York: Plenum Press.

Shriner, D.S., C.R. Richmond, and S.E. Lindberg (1980) Atmospheric Sulfur Deposition—Environmental Impact and Health Effects. Ann Arbor, MI: Ann Arbor Science.

Shugart, H.H., S.B. McLaughlin, and D.C. West (1980) Forest models: Their development and potential applications for air pollution effects research. Pages 203-214, Effects of Air Pollutants on Mediterranean and Temperate Forest Ecosystems. Proc. of a Symposium, June 22-27, 1980, Riverside, CA. General Technical Report PSW-43. Washington, DC: Forest Servic, USDA.

Shum, V.S., and W.O. Loveland (1974) Atmospheric trace element concentrations associated with agricultural field burning in the Willamette Valley of Oregon. Atmos. Environ. 8:645-655.

Sickles, J.E., W.C. Eaton, L.A. Ripperton, and R.S. Wright (1977) Literature Survey of Emissions Associated with Emerging Energizing Technologies. EPA 600/7-77-104. Washington, DC: U.S. EPA.

Simoneit, B.R.T. (1979) Biogenic lipids in eolian particulates collected over the ocean. Pages 233-244, Carbonaceous Particles in the Atmosphere, edited by T. Novakov. LBL-9037. Berkeley, CA: National Science Foundation/Lawrence Radiation Laboratory.

Simoneit, B.R.T., and M.A. Mazurek (In press) Air pollution: The organic components. Crit. Rev. Environ. Control.

Singer, C. (1959) A History of Biology. 3rd ed. New York: Abelard-Schuman.

Singh, B.R. (1980) Effects of acid precipitation on soil and forest. 3. Sulfate sorption by acid forest soil. Pages 194-195, Ecological Impact of Acid Precipitation. Proceedings of an International Conference, Sandefjord, Norway, March 11-14, 1980, edited by D. Drablös and A. Tollan. Oslo-Aas, Norway: SNSF project.

Sivard, R.L. (1979) Trends in World Energy Sources. World Energy Survey. Leesburg, VA: World Priorities.

Skei, J., and P.E. Paus (1979) Surface metal enrichment and partioning of metals in a dated sediment core from a Norwegian fjord. Geochim. Cosmochim. Acta. 43:239-246.

Skye, E. (1968) Lichens and air pollution. Acta Phytogeogr. Suecica No. 52.

Slinn, W.G.N., L. Hasse, B.B. Hicks, A.W. Hogan, D. Lal, P.A. Liss, K.O. Munnich, G.A. Sehmel, and O. Vittori (1978) Some aspects of the transfer of atmospheric trace constituents past the air-sea interface. Atmos. Environ. 12:2055-2087.

Smith, D.M., J.J. Griffin, and E.D. Goldberg (1973) Elemental carbon in marine sediments: A baseline for burning. Nature 241:268-270.

Smith, K.A., J.M. Bremner, and M.A. Tabatabai (1973) Sorption of gaseous atmospheric pollutants by soils. Soil Sci. 116:313-319.

Smith, R.A. (1852) On the air and rain of Manchester. Mem. Lit. Philos. Soc. Manchester, Series 2. 10:207-217.

Smith, R.A. (1872) Air and Rain. London: Longmans, Green.

Snaydon, R.W. (1962) The growth and competitive ability of contrasting natural populations of Trifolium repens L. on calcareous and acid soils. J. Ecol. 50:439-447.

Snieszko, S.F. (1974) The effects of environmental stress on outbreaks of infectious diseases of fish. J. Fish. Biol. 6:197-208.

Söderlund, R., and B.H. Svensson (1976) The global nitrogen cycle. Ecological Bulletin 22. Stockholm: Swedish National Research Council.

Sonstegard, R.A., and J.F. Leatherland (1980) Aquatic organism pathobiology as a sentinel system to monitor environmental carcinogens. Pages 513-520, Hydrocarbons and Halogenated Hydrocarbons in the Aquatic Environment, edited by B.K. Afgan and D. Mackay. New York: Plenum Press.

Sorensen, J. (1978) Capacity of denitrification and reduction of nitrate to ammonia in a coastal marine sediment. Appl. Environ. Microbiol. 35:301-305.

Springer, K.J. (1978) Exhaust particulate--The diesel's Achilles heel. 71st Annual Meeting, June 25-30, 1978, Air Pollution Control Association, Houston, TX. Pittsburgh, PA: Air Pollution Control Association.

Spry, D.J., C.M. Wood, and J. Hodson (1981) The effects of environmental acid on freshwater fish with particular reference to the softwater lakes in Ontario and modifying effects of heavy metals in the environment. Can. Tech. Rept., Fisheries and Aquat. Sci. No. 999. Burlington, Ontario: Dept. Fisheries and Oceans.

Steemann, N.E., and S. Wium-Andersen (1970) Copper ions as poison in the sea and in freshwater. Marine Biol. 6:93-97.

Stewart, W.D.P., G.P. Fitzgerald, and R.H. Burris (1967) In situ studies on nitrogen fixation using the acetylene reduction technique. Proc. Natl. Acad. Sci. USA 58:2071-2078.

Stocks, P., B.T. Commins, and K.V. Aubrey (1961) A study of polycyclic hydrocarbons and trace elements in Merseyside and other northern localities. Int. J. Air Water Pollut. 4:141-153.

Stokes, P.M., T.C. Hutchinson, and K. Krauter (1973) Heavy metal tolerance in algae isolated from contaminated lakes near Sudbury, Ontario. Water Pollut. Res. Can. 8:178-187.

Stokes, P.M. (1975) Adaption of green algae to high levels of copper and nickel in aquatic environments. Pages 135-154, International Conference on Heavy Metals in the Environment, Vol. 2: Pathways, edited by T.C. Hutchinson. Toronto: University of Toronto.

Strand, L. (1980) The effect of acid precipitation on tree growth. Pages 64-67, Ecological Impact of Acid Precipitation. Proceedings of an International Conference, Sandefjord, Norway, March 11-14, 1980, edited by D. Drablös and A. Tollan. Oslo-Aas, Norway: SNSF project.

Strojan, C.L. (1980) Forest litter decomposition in the vicinity of a zinc smelter. Oecologia 32:203-212.

Stuanes, A.O. (1980) Effects of acid precipitation on soil and forest. 5. Release and loss of nutrients from a Norwegian forest soil due to artificial rain of varying acidity. Pages 198-199, Ecological Impact of Acid Precipitation. Proceedings of an International Conference, Sandefjord, Norway, March 11-14, 1980, edited by D. Drablös and A. Tollan. Oslo-Aas, Norway: SNSF project.

Swain, R.E. (1949) Smoke and fume investigations. A historical review. Ind. Eng. Chem. 41:2384-2388.

Swarts, F.A., W.A. Dunson, and J.E. Wright (1978) Genetic and environmental factors involved in increased resistance of brook trout to sulphuric acid solutions and mine acid polluted waters. Trans. Amer. Fish. Soc. 107:651-677.

Talbot, V., and R.J. Magee (1978) Naturally-occurring heavy metal binding proteins in invertebrates. Archives Environmental Contamination and Toxicol. 7:73-81.

Tallis, J.H. (1964) Studies on southern Pennine peats. III. The behavior of Sphagnum. J. Ecol. 52:345-353.

Tamm, C.O. (1953) Growth, yield and nutrition in carpets of a forest moss (Hylocomium splendens). Medd. Statens Skogsforskningsinst 43, No. 1.

Tamm, C.O., and T. Troedsson (1957) A new method for the study of water movement in soil. Geol. Foeren. Stockholm Foerh. 79:581-587.

Tamm, C.O. (1976) Acid precipitation: Biological effects in soil and on forest vegetation. AMBIO 5:235-238.

Tamm, C.O. (1977) Skogsmarkens forsurning, orsaker och motatgarder. Sveriges Skogvardsforbunds Tidskr. 75:189-200.

Tamm, C.O., and G. Wiklander (1980) Effects of artificial acidification with sulfuric acid on tree growth in Scots pine forest. Pages 188-189, Ecological Impact of Acid Precipitation. Proceedings of an International Conference, Sandefjord, Norway, March 11-14, 1980, edited by D. Drablös and A. Tollan. Oslo-Aas, Norway: SNSF Project.

Tandon, S.K., L. Magos, and J.R.P. Cabral (1980) Protection against mercuric chloride by nephrotoxic agents which do not induce thionein. Toxicol. and Appl. Pharm. 52:227-236.

Tatsuyamo, K., H. Egawa, H. Yamamoto, and H. Senmaru (1975) Tolerance of cadmium-resistant microorganisms to other metals. Trans. Mycol. Soc. Jpn. 16:79-85.

Taylor, G.E. (1978) Genetic analysis of ecotypic differentiation within an annual plant species, Geranium carolinianum L., in response to sulfur dioxide. Bot. Gaz. 139:362-368.

Taylor, G.R. (1963) The Science of Life. New York: McGraw-Hill.

Thomas, T.L., P. Decoufle, and R. Moure-Eraso (1980) Mortality among workers employed in petrochemical refining and petrochemical plants. J. Occup. Med. 22:97-103.

Thomas, W.A., ed. (1972) Indicators of Environmental Quality. New York: Plenum Publishing Corporation.

Thompson, M.E., F.C. Elder, A.R. Davis, and S. Whitlow (1980) Evidence of acidification of rivers of Eastern Canada. Pages 244-245, Ecological Impact of Acid Precipitation. Proceedings of an International Conference, Sandefjord, Norway, March 11-14, 1980, edited by D. Drablös and A. Tollan. Oslo-Aas, Norway: SNSF project.

TIE, The Institute of Ecology (1977) Experimental Ecological Reserves, A Proposed National Network. Washington, DC: U.S. Government Printing Offfice.

Tirén, T., J. Thorin, and H. Nômmik (1976) Denitrification measurements in lakes. Acta Agric. Scand. 26:175-184.

Toivonen, P.M.A., and G. Hofstra (1979) The interaction of copper and sulfur dioxide in plant injury. Can. J. Plant Sci. 59:475-479.

Tomlinson, G.H., R.J.P. Brouzes, R.A.N. McLean, and J. Kadeck (1980) The role of clouds in atmospheric transport of mercury and other pollutants. Pages 134-137, Ecological Impact of Acid Precipitation. Proceedings of an International Conference, Sandefjord, Norway, March 11-14, 1980, edited by D. Drablös and A. Tollan. Oslo-Aas, Norway: SNSF project.

Tomlinson, G.H. (1981) Acid rain and the forest--the effect of
    aluminum and the German experience. DOMTAR Publ. No. 74-7124-13.
    Senneville, Quebec: DOMTAR Research Centre.

Traaen, T.S. (1980) Effects of acidity on decomposition of organic
    matter in aquatic environments. Pages 340-341, Ecological Impact
    of Acid Precipitation. Proceedings of an International
    Conference, Sandefjord, Norway, March 11-14, 1980, edited by
    D. Drablös and A. Tollan. Oslo-Aas, Norway: SNSF project.

Troedsson, T. (1980) Ten years acidification of Swedish forest soils.
    Page 184, Ecological Impact of Acid Precipitation. Proceedings of
    an International Conference, Sandefjord, Norway, March 11-14,
    1980, edited by D. Drablös and A. Tollan. Oslo-Aas, Norway: SNSF
    project.

Trojnar, J.R. (1977a) Egg and larval survival of white suckers
    (Catostomus commersoni) at low pH. J. Fish. Res. Bd. Can.
    34:262-266.

Trojnar, J.R. (1977b) Egg hatchability and tolerance of brook trout
    (Salvelinus fontinalis) fry at low pH. J. Fish. Res. Bd. Can.
    34:574-579.

Troutman, D.E., and N.E. Peters (1980) Comparison of lead, manganese,
    and zinc transport in three Adirondack lake watersheds, New York.
    Pages 262-263, Ecological Impact of Acid Precipitation.
    Proceedings of an International Conference, Sandefjord, Norway,
    March 11-14, 1980, edited by D. Drablös and A. Tollan. Oslo-Aas,
    Norway: SNSF project.

Tusneem, M.E., and W.H. Patrick, Jr. (1971) Nitrogen transformations
    in waterlogged soil. Bulletin 657, Department of Agronomy. Baton
    Rouge, LA: Louisiana State University.

Tveite, B. (1980) Effects of acid precipitation on soil and forest.
    8. Foliar nutrient concentrations in field experiments. Pages
    204-205, Ecological Impact of Acid Precipitation. Proceedings of
    an International Conference, Sandefjord, Norway, March 11-14,
    1980, edited by D. Drablös and A. Tollan. Oslo-Aas, Norway: SNSF
    project.

Tyler, G. (1974) Heavy metal pollution and soil enzymatic activity.
    Plant Soil 41:303-311.

Tyler, G. (1976a) Heavy metal pollution, phosphatase activity and
    mineralization of organic phosphorus in forest soils. Soil Biol.
    Biochem. 8:327-332.

Tyler, G. (1976b) Influence of vanadium on soil phosphatase activity.
    J. Environ. Qual. 5:216-217.

Tyler, G. (1978) Leaching rates of heavy metal ions in forest soil.
    Water, Air, Soil Pollut. 9:137-148.

Ulrich, R., R. Mayer, and P.K. Khanna (1980) Chemical changes due to
    acid precipitation in a loess-derived soil in central Europe.
    Soil Sci. 130:193-199.

United Nations (1976) World Energy Supplies 1950-1974. New York.

United Nations (1978) World Energy Supplies 1972-1976. New York.

United States-Canada Research Consultation Group on the Long-Range
    Transport of Air Pollutants (1979) The LRTAP Problem in North
    America: A Preliminary Overview. Downsview, ON: Information
    Directorate, Atmospheric Environment Service.

U.S. Department of Health, Education, and Welfare (1979) Smoking and Health. A Report of the Surgeon General. DHEW Publication No. PHS 79-50066. Washington, DC: U.S. Government Printing Office.

U.S. Environmental Protection Agency (1976) National Air Quality and Emission Trends Report, 1975. EPA-450/1-76-002. Research Triangle Park, NC: U.S. EPA.

U.S. Environmental Protection Agency (1977) Compilation of Air Pollutant Emission Factors. AP-42. Office of Air and Waste Management. Research Triangle Park, NC: U.S. EPA.

U.S. Environmental Protection Agency (1978a) National Air Quality, Monitoring and Emissions Report, 1977. EPA-450/2-78-052. Research Triangle Park, NC: U.S. EPA.

U.S. Environmental Protection Agency (1978b) Preliminary Assessment of the Sources, Control and Population Exposure to Airborne Polycyclic Organic Matter as Indicated by Benzo(a)pyrene. Prepared by Energy and Environmental Analysis Inc., Contract No. 68-02-3836. Research Triangle Park, NC: U.S. EPA, Office of Air Quality Planning and Standards.

U.S. Environmental Protection Agency (1980) Controlling Nitrogen Oxides. EPA-600/8-80-004. Research Triangle Park, NC: U.S. EPA.

U.S. Geological Survey (1970) Mercury in the Environment. U.S. Geological Survey Professional Paper No. 713. Washington, DC: U.S. Department of the Interior.

U.S. Senate, Committee on Government Affairs (1979) Carbon Dioxide Accumulation in the Atmosphere, Synthetic Fuels and Energy Policy. Washington, DC: U.S. Government Printing Office.

Vallentyne, J.R. (1974) The Algal Bowl, Lakes and Man. Miscellaneous Special Publication No. 22. Ottawa: Fisheries Research Board of Canada.

Van Dam, H., G. Suurmond, and C. TerBraak (1980) Impact of acid precipitation on diatoms and chemistry of Dutch moorland pools. Pages 298-299, Ecological Impact of Acid Precipitation. Proceedings of an International Conference, Sandefjord, Norway, March 11-14, 1980, edited by D. Drablös and A. Tollan. Oslo-Aas, Norway: SNSF project.

Vanderhoef, L.N., C.Y. Huang, R. Musil, and J. Williams (1974) Nitrogen fixation (acetylene reduction) by phytoplankton in Green Bay, Lake Michigan, in relation to nutrient concentrations. Limnol. Oceanogr. 19:119-125.

Vangenechten, J.H.D., and O.L.J. Vanderborght (1980) Acidification of Belgian moorland pools by acid sulfur-rich rain water. Pages 246-247, Ecological Impact of Acid Precipitation. Proceedings of an International Conference, Sandefjord, Norway, march 11-14, 1980, edited by D. Drablös and A. Tollan. Oslo-Aas, Norway: SNSF project.

Walker, N., and K.N. Wickramasinghe (1979) Nitrification and autotrophic nitrifying bacteria in acid tea soils. Soil Biol. Biochem. 11:231-236.

Ward, N.I., and R.R. Brooks (1978) Lead levels in sheep organs resulting from pollution from automotive exhaust. Environ. Pollut. 17:7-12.

Warwick, W. (1980a) Chironomidae (Diptera): Responses to 2800 years of cultural influence; A paleolimnological study with special reference to sedimentation, eutrophication and contamination processes. Can. Entomol. 112:1193-1238.

Warwick, W. (1980b) Pasqua Lake, southeastern Saskatchewan: A preliminary assessment of trophic status and contamination based on the Chironomidae (Diptera). Pages 255-267, Chironomidae: Ecology, Systematics, Cytology and Physiology, edited by D.A. Murray. Oxford, England: Pergamon Press.

Watson, A.P., R.I. van Hook, D.R. Jackson, and D.E. Reichle (1976) Impact of a lead mining-smelting complex on the forest floor litter arthropod fauna in the New Lead Belt Region of Missouri. Publication No. 881. Oak Ridge, TN: Oak Ridge National Laboratory, Environmental Sciences Division.

Watson, H.C. (1833) Observations on the affinities between plants and subjacent rocks. Mag. Nat. Hist. 6:424-427.

Watt, W.D., D. Scott, and S. Ray (1979) Acidification and other chemical changes in Halifax County lakes after 21 years. Limnol. Oceanogr. 24(6):1154-1161.

Weaver, J.E., and F.E. Clements (1938) Plant Ecology. 2nd ed. New York: McGraw Hill.

Welch, H.E., J.W.M. Rudd, and D.W. Schindler (1980) Methane addition to an arctic lake in winter. Limnol. Oceanogr. 25:100-113.

Wells, R.M., and J.D. Yarbrough (1972) Retention of ^{14}C-DDT in cellular fractions of vertebrate insecticide-resistant and susceptible fish. Toxicol. Appl. Pharmocol. 22:409-414.

Went, F. (1960) Organic matter in the atmosphere, and its possible relation to petroleum formation. Proc. Natl. Acad. Sci. USA 46:212-221.

West, D.C., S.B. McLaughlin, and H.H. Shugart (1980) Simulated forest response to chronic air pollution stress. J. Environ. Qual. 9:43-49.

Whitby, L.M., and T.C. Hutchinson (1974) Heavy metal pollution in the Sudbury mining and smelting region of Canada. II. Soil toxicity tests. Environ. Conservation 1:191-200.

White, R.K., E.P. Taiganides, and G.D. Cole (1971) Chromatographic identification of malodors from dairy animal waste. Pages 110-113, Livestock Waste Management and Pollution Abatement. St. Joseph, MI: American Society of Agricultural Engineers.

Wijler, J., and C.C. Delwiche (1954) Investigations on the denitrification process in soil. Plant Soil 5:155-169.

Wiklander, L. (1973) The acidification of soil by acid precipitation. Grundforbattring 26:155-164.

Wiklander, L. (1974) Leaching of plant nutrients in soils. 1. General principles. Acta Agric. Scand. 24:349-356.

Wiklander, L. (1975) The roles of neutral salts in the ion exchange between acid precipitation and soils. Geoderma 14:93-105.

Wiklander, L. (1979) Leaching and acidification of soils. Section 4.3, Ecological Effects of Acid Precipitation. EPRI Report No. EA-79-6-LD. Palo Alto, CA: Electric Power Research Institute.

Wiklander, L. (1980) Interactions between cations and anions influencing absorption and leaching. Pages 239-254, Effects of

Acid Precipitation on Terrestrial Ecosystems, edited by T.C. Hutchinson and M. Havas. New York: Plenum Press.

Wilkniss, P.E., J.W. Swinnerton, R.A. Lamontagne, and D.J. Bressan (1975) Trichlorofluoromethane in the troposphere, distribution and increase, 1971-1974. Science 187:832-833.

Williams, S.T., T. McNeilly, and E.M.H. Wellington (1977) The decomposition of vegetation growing on metal mine waste. Soil Biol. Biochem. 9:271-275.

Wilson, L.G. (1960) The transformation of ancient concepts of respiration in the seventeenth century. Isis 51:161-172.

Wilson, W.E. (1978) Sulfates in the atmosphere: A progress report on Project MISTT. Atmos. Environ. 12:537-547.

Windom, H., J. Griffin, and E.D. Goldberg (1967) Talc in atmospheric dusts. Environ. Sci. Technol. 1:923-926.

Windsor, J.G., Jr., and R.A. Hites (1979) Polycyclic aromatic hydrocarbons in Gulf of Maine soils. Geochem. Cosmochem. Acta 43:27-33.

Winge, D.R., R. Premakumar, and K.V. Rajagopalan (1975) Metal-induced formation of metallothionein in rat liver. Arch. Biochim. Biophys. 170:242-252.

Witting, M. (1947) Kajonbestämningar i myrvatten. Botaniska Notiser Lund, pp. 287-304.

Witting, M. (1948) Fortsatta katjonbestämningar i myrvatten sommaren 1947. Sven. Botanisk Tidskr. 42:116-134.

Wood, C.W., and T.N. Nash (1976) Copper smelter effluent effects on Sonoran desert vegetation. Ecology 57:1311-1316.

Wood, J.M. (1974) Biological cycles for toxic elements in the environment. Science 183:1049-1052.

Wood, J.M., and E.D. Goldberg (1977) Impact of metals on the biosphere. Pages 137-153, Global Chemical Cycles and Their Alterations by Man, edited by W. Stumm. Physical and Chemical Sciences Research Report 2. West Berlin: Dahlem Konferenze.

Woodward, J. (1699) Some thoughts and experiments concerning vegetation. Philos. Trans. Royal Soc. Lond. 21:193-227.

Woodwell, G.M., and A.H. Sparrow (1963) Predicted and observed effects of chronic gamma radiation on a near-climax forest ecosystem. Radiat. Botany 3:231-237.

Woodwell, G.M. (1967) Radiation and the patterns of nature. Science 156:461-470.

Woodwell, G.M. (1970) Effects of pollution on the structure and function of ecosystems. Science 168:429-433.

Woolson, E.A. (1977) Generation of alkylarsines from soil. Weed Sci. 25:412-416.

Wong, C.S. (1978) Atmospheric input of carbon dioxide from burning wood. Science 200:197-200.

Wong, P.T.S., Y.K. Chau, and P.L. Luxon (1978) Toxicity of a mixture of metals on freshwater algae. J. Fish. Res. Bd. Can. 35:479-481.

World Health Organization (1961) Air Pollution. WHO Monograph Series, No. 46. Geneva: World Health Organization.

Wright, R.F., T. Dale, T.E. Gjessing, G.R. Hendry, A. Henriksen, M. Johannessen, and I.P. Muniz (1976) Impact of acid precipitation on freshwater ecosystems in Norway. Water, Air, Soil Pollut. 6:483-499.

Wright, R.F., N. Conroy, W.T. Dickson, R. Harriman, A. Henriksen, and C.L. Schofield (1980) Acidified lake districts of the world: A comparison of water chemistry of lakes in southern Norway, southern Sweden, southwestern Scotland, the Adirondack Mountains of New York, and southeastern Ontario. Pages 377-379, Ecological Impact of Acid Precipitation. Proceedings of an International Conference, Sandefjord, Norway, March 11-14, 1980, edited by D. Drablös and A. Tollan. Oslo-Aas, Norway: SNSF project.

Yan, N.D., and P.M. Stokes (1978) Phytoplankton of an acidic lake, and its responses to an experimental alteration of pH. Environ. Conservation 5:93-100.

Yeager, K.E. (1979) Cleaning up coal. Electric Power Res. Inst. J. 4:2-3.

Yee, M.S., H.L. Bohn, and S. Miyamoto (1975) Sorption of sulfur dioxide by calcareous soils. Soil Sci. Soc. Am. Proc. 39:268-270.

Yoshida, T., and M. Alexander (1970) Nitrous oxide formation by Nitrosomonas europaea and heterotrophic microorganisms. Soil Sci. Soc. Am. Proc. 34:880-882.

Yoshikawa, H. (1970) Preventitive effect of pretreatment with low dose metals on the acute toxicity of metals in mice. Industrial Health 8:184.

Young, R.J., N.C. Dondero, D.C. Ludington, and R.C. Loehr (1971) Poultry waste management and the control of associated odors. Pages 98-104, Identification and Measurement of Environmental Pollutants, edited by B. Westley. Ottawa: Natural Resources Council of Canada.

Zimmerman, P.R., R.B. Chatfield, J. Fishman, P.J. Crutzen, and P.L. Hanst (1978) Estimates on the production of CO and H_2 from the oxidation of hydrocarbons emissions from vegetation. Geophys. Res. Lett. 5:679-682.

INDEX

258